A HANDBOOK OF BIOLOGICAL INVESTIGATION

SIXTH EDITION

Harrison W. Ambrose III, Ph.D.[1]
Katharine Peckham Ambrose, M.S., J.D.[1]
Douglas J. Emlen, Ph.D.[2]
Kerry L. Bright, Ph.D.[2]

⊞ Hunter Textbooks Inc.

[1] 6507 Bull Run Road,
Knoxville, Tennessee 37938

[2] Division of Biological Sciences,
University of Montana, Missoula, MT 59812

Hunter Textbooks Inc.
701 Shallowford Street
Winston-Salem, NC 27101

E-MAIL US AT **hunter@rbdc.com**
PURCHASE THIS BOOK THROUGH OUR WEB SITE AT **www.huntertextbooks.com**

Dedication

This book is dedicated to J. Hill Hamon and to Donald Y. Young, remarkable teachers. When I was an undergraduate student, I was fortunate enough to have J. Hill as one of my instructors. His dedication to teaching and his willingness to help the individual student not only helped me make my career choices but has always served as my example.

Later, as a faculty member, struggling to introduce and test the investigatory laboratory teaching technique, I was privileged to have the help and support of another such teaching assistant Donald Y. Young. Don, too, has the boundless energy, patience and skill needed to help another generation of biology students.

H.W.A.III

Acknowledgments

The authors gratefully acknowledge their debt to Diane Schmidt, biology librarian at the University of Illinois, for her generous contributions of Chapter Nine, Using the Library and Chapter Ten, An introduction to Biological Literature. Nicholas J. Cheper and Donald Y. Young generously helped with Chapter Eight.

In addition we would like to thank John Christy, Steve Kessell and Don Riker who, as teaching assistants, contributed their suggestions, criticisms and creative energy toward our development of the investigatory laboratory technique.

We would also like to thank the many other dedicated teaching assistants at Cornell University, University of Illinois and University of Tennessee at Knoxville, who, like those named above, were willing to give more than was required.

Finally, we thank our son and daughter-in-law, Douglas J. Emlen and Kerry L. Bright for contributing the wisdom and energy of a new generation to the re-generation of this book.

H.W.A.III and K.P.A.

Contents

Preface

This handbook is neither a text nor a laboratory manual. It is written as a reference aid to be used by biology students at any time they are first exposed to the scientific method and to original research. It guides the beginner through elementary experimental design, data collection and analysis, and the subsequent writing of a scientific report. In addition, it provides an introduction to the literature of biology and a basic search strategy for finding relevant, published information in a library.

We anticipate that this book will be used with a text in a particular subject specialty, a manual of laboratory techniques, a guide to field collection, or with a source book on the care and feeding of laboratory plants and animals. None of these books alone equips the undergraduate or beginning graduate student with the potential for making scientific discoveries. The supplementary use of this handbook, however, can add the excitement of original student research to any course in the life sciences.

At the Universities of Tennessee and Illinois, this material has been used in sections of introductory biology taught in the investigatory laboratory mode – a method of teaching which prepares and later permits students to design, pursue, analyze, and report on independent research. The students were never asked to learn the material in this book but merely to use it as they would a dictionary or other reference. No mathematical ability was required; laboratories were equipped with calculators for student use.

Portions of this book have also been successfully introduced at advanced undergraduate and beginning graduate levels. No matter when a student begins original scientific experimentation, a careful, gradual introduction is required.

The enterprise is doomed if:

- students are not adequately prepared before they begin independent work;
- students are not given appropriate credit for the actual experimental effort (as well as the eventual scientific report);
- the instructor insists that experiments be truly original and/or successful;
- the instructor insists that the final reports be of publishable quality; and if
- the instructor believes the students incapable of high-quality, creative, independent work.

For those interested in how freshman might be exposed to independent research, we have included a general outline of the first crucial weeks in the laboratory (see the Appendix to this handbook). For more information on the concept of the investigatory laboratory, we strongly recommend the seminal report entitled **"The Laboratory: A Place to Investigate"** prepared by the Commission on Undergraduate Education in the Biological Sciences in 1972.

CHAPTER ONE

Introduction: What Is Science?

The whole of science is nothing more than a refinement of everyday thinking.

— Albert Einstein, *Physics and Reality*, 1936

S cience is a process—a way of thinking about problems and a methodology for solving problems. Scientists use an analytical method involving observation, experimentation, and both inductive and deductive logic. This well-defined method distinguishes science from other disciplines such as philosophy or theology.

Each of these disciplines (science, philosophy and theology), however, may be used to ask questions and acquire knowledge. For example, if the fern in your dormitory room dies unexpectedly, you may ask, "Why did God kill my fern?" Your question may be answered by revelation through prayer or meditation, but this method of acquiring information lies outside the scope of science. Science simply cannot address this question because **science is limited to the study of the physical universe, the natural world.** Explanations that require supernatural interventions are outside the realm of science.

Let us take a scientific approach to understanding the death of your fern. The process of acquiring information through the scientific method involves the postulating and testing of hypotheses. **A hypothesis is a possible explanation for an observed phenomenon** (see Box 1.1). In this case, you ask the question, "What conditions led to the death of my fern?" Suppose you hypothesize that your plant died because it did not receive enough water. This would be a good guess because you already know water is essential for plant life. Maybe you fear you let a week go by without watering it, and that insufficient water has caused its demise. Your roommate says, however, that you did not need to water it more than once a week anyway. You are sure

1

that watering only once a week is nowhere near enough. You predict that *if* you bought another fern and watered it only once a week, *then* it too would die. **Scientists use deductive (if . . . then) logic to make and then test predictions on the basis of their hypotheses**. In a sense, you have now made a prediction on the basis of your hypothesis.

Let us restate your prediction in this way: "If I buy another fern of the same kind, in the same kind of soil and container, and house it in the same part of my room, and if I water it only once a week, then it will die." Your roommate says, "Rubbish!" You are curious enough to go out and buy another fern. (It is the possession of a curious mind that unites scientists everywhere no matter how diverse their backgrounds, interests and education.)

You perform your experiment and, after two weeks, your second fern dies. "I told you so! Watering only once a week is not enough," you inform your roommate.

"You must have done something different with it this time," he retorts.

"No, everything was the same as before—same kind of plant, same place in the room, same everything."

Although neither of you ever needs to consciously think these thoughts, you are both making some of the basic assumptions that all scientists make. You are assuming **that there is order in the universe, that the human mind is capable of comprehending it, and that if conditions are the same, the results will be the same.** All scientific experiments are based on the assumption that similar results would occur if the experiment were repeated under similar circumstances.

Let us look at your experiment again. Did you prove anything? You may think you have. Your experimental results support your hypothesis, but your roommate remains unconvinced. One of the hazards of deductive logic is that both true and false hypotheses may give rise to true predictions, and additional evidence is required in these instances to discriminate among the possible hypotheses.

A noteworthy example of this is the hypothesis that the sun revolves around the earth (geocentrism). This *false* hypothesis gave rise to *true* predictions about sunrise and sunset that were convincing enough to make this explanation appear to be true. For more than a thousand years astronomers accepted this false hypothesis. Subsequent evidence—data on planetary orbits, especially the movements of Mars—allowed astronomers to reject the false hypothesis that the sun revolved around the earth and adopt the presently accepted view that the earth, and the other planets, revolve around the sun (heliocentrism).

Back to your ferns. You may now be convinced that inadequate watering has caused the death of two plants. Your roommate suggests that you watered them both too much. He is even willing to sponsor some additional research and brings home three more ferns. You hypothesize that *if* you water your plants every day (all other conditions being the same) *then* they will live. You choose one plant to be kept in wet soil by watering every day, one to be watered every other day, and one (to satisfy your roommate) to be watered every 10 days. What if all the plants die? Now have you shown anything? This time you have. You have failed to support your hypothesis that inadequate watering was killing the ferns. You may be discouraged, but actually you have succeeded in rejecting your hypothesis. If you can reject a hypothesis, you have taken a step forward. **All science advances**

3

by the rejection of hypotheses. There is no such thing as proof. Each step forward involves disproof (see Box 1.2).

You and your roommate may still be interested in testing ferns in an attempt to determine the cause of this high mortality rate. You may well be out of money and have only your curiosity and enthusiasm left. This is the normal condition of a scientist. You may want to apply for a grant.

Why would anybody care about your specific ferns? Well, probably nobody does. However, **good scientific research produces results from which you can generalize.** You may be able to establish something about ferns in general. Perhaps you approach your botany department, explain what has happened to all your ferns, and request a small grant to cover the purchase of additional ferns. If they see anything interesting about the death of your ferns, they may be willing to help you, provided that you supply them with a research proposal.

They will point out that water does not seem to be a factor, but have you considered that the plants may not have had enough sunlight? (In fact your plants were on a bookshelf in a rather dark corner of the room.) Or that the room temperature might have been inappropriate? (In fact, because of an energy shortage, your dormitory has been maintained at 60° F each night and it is not unreasonable to suppose that this might be too cold for a fern.) Or that, after a certain age, the ferns are too large for the pots they are sold in and need repotting? If you are ever to establish the cause of death of your plants, you will have to examine each of these possibilities—and there may well be others you have not yet thought of. Maybe somebody in your dorm is using your plants as an ashtray, for example.

There is a very well established method of attack on a problem such as yours. The two essential first steps are a library search to find out what is already known (there may well be a book all about the care and feeding of ferns) and a brainstorming session. In the latter, you and your friends should **list all the possible explanations you can think of**.

Each explanation should be written down and turned into an explicit hypothesis. By making predictions (*if* I do this, *then* that will happen), you should be able to figure out what experiment would disprove each hypothesis. If you hypothesized that putting the fern in the south-facing window would allow it to live (and your library reading has suggested that this is normally appropriate), then it is easy to see that putting some ferns in the window will be part of the next experiment. If these live, you have proven nothing, but you may be able to infer that sunlight was the problem all along. If they die, however, you can check off the too-little-sunlight hypothesis from your list (i.e., you can reject this hypothesis as an explanation for the death of your fern). It was clearly not the factor causing death.

You should work out an experimental design in which only one variable factor is examined at a time while all other factors are held constant. Perhaps three plants will be placed on the windowsill and three will be put on the dark side of the room but warmed by a space heater (with precautions that the heater not provide additional light). Three other plants may be purchased and repotted, but kept on the bookshelf as before. If your literature agrees with this formula, all the plants will be watered once a week. In each experiment, the hope is to manipulate only one variable at a time. Certain experimental subjects are normally held as controls—in this experiment that would mean some additional plants experienced no manipulation of light, heat, or pot.

If you actually performed this experiment and all of your plants died, each of your variables would have been eliminated as the causative factor, all hypotheses rejected, and you would be back at the drawing board, thinking up new hypotheses. This is not an unusual occurrence. Other factors would then have to be explored such as manipulations with the soil. Let us suppose, however, that you reject all your hypotheses but the too-little-sunlight one (only the plants on the windowsill survived). Would you have *proven* inadequate sunlight caused the death of all your other plants? No, but you would have grounds for a strong inference that sunlight was the factor (i.e., the experiment has *supported* this hypothesis). The more ferns you see surviving in south-facing windows, the stronger would be your inference.

As we have said before, there is no such thing as proof in science. There is only disproof. For any hypothesis you think you have established as the correct explanation, there is always the possibility of an experiment that would disprove it. Whenever you design an experiment to test a hypothesis, try to think of the experiment that would disprove it. If you fail to disprove it, you can infer **(but never absolutely prove)** that your hypothesis was correct (see Box 1.2).

Many rather subtle points have been touched on in this introduction. Each will be amplified and clarified during the course of this book. We hope that this hypothetical situation has shown you several things. First of all, any phenomenon not

---

Box 1.2   Proof in Science

It is common to hear on television or radio: "Science has now proven..." or "Scientists prove...." Such bold statements make effective sound bites promoting commercial products. Claims about everything from detergents to medicines sound more believable if they are backed by scientific "proof." However, this illustrates an important and unfortunate misconception about the nature of science. Science never proves anything. **Knowledge advances by the systematic rejection of available hypotheses**. We are limited by the list of hypotheses we generate. Since it is always possible that the true explanation was not included in the list, scientific explanations, by definition, are tentative, always subject to modification in the light of new information. The danger of this misconception about proof is that the general public grows to view science as a collection of **facts**, rather than as a **process**. When a "proven" scientific hypothesis is rejected, people grow to distrust science. Ironically, these step-wise advances in our understanding provide the best illustrations of the self-correcting nature, and thus the power, of scientific explanations.

---

immediately explicable (such as the sudden death of a fern) might stimulate a small-scale scientific experiment. Experiments frequently lead to new questions. Scientists are often led from question to question, gradually checking off rejected hypotheses as if they were detectives solving a mystery. In fact, a scientist is a detective and nature is filled with mysteries. Adequate reading and preparation can save a great deal of time and effort; and a very careful plan of attack, in which variable factors are segregated and examined one at a time, can provide a rigorous and generalizable explanation for any natural phenomenon.

CHAPTER TWO
Asking Questions and Formulating Hypotheses

Scientific information is gathered by asking questions. Not all questions are susceptible to scientific inquiry, as we have said before. Questions which do not deal with the physical world (What is truth? Is there a God? Is it moral to "mercy kill" a person suffering from a painful, terminal disease? etc.) are outside the scope of science. Science cannot, by definition, address these types of questions.

Extremely broad questions about physical phenomena are also inappropriate for our purposes even though they fall within the scope of science. They are not feasible for a beginning researcher. For example, asking "What is sunlight?" would be too broad a question, but asking whether sunlight were required for the germination of a seed would be entirely appropriate. Or asking "What is gravity?" would be too broad for the beginner to get an experimental handle on it. However, a question about the specific effect of gravity on plant growth would be entirely suitable for study. It is important for any scientist to learn how to ask focused, and therefore feasible, research questions.

Before you are ready to select a topic for scientific research, you must practice asking questions about the physical world and figuring out what kind of information would be needed to answer these questions. Those of us fortunate enough not to have had our curiosity stifled by exasperated parents or teachers are constantly asking ourselves questions. We wonder when the robins will return, whether the coat of the woolly bear caterpillar really can be used to predict the severity of winter weather, and why the male cardinal is more brightly colored than his mate? If, on the other hand, our childhood questions were frequently answered with the uninformative, "Because that's the way things are," we may have lost our natural curiosity. We may find we think of no questions at all.

To rekindle the spark, we must now exercise our minds, consciously looking for questions and trying to think of ways to answer them. This process is one of the most exciting aspects of a scientific endeavor, involving a mixture of observation, creativity and reasoning. All questions start with an observation. We illustrate this with an example. While cross-country skiing through a woodland, you notice several insects crawling on the snow. You take off your mittens and pick up several to look at more closely. Within a minute of picking them up, you notice that all the insects die. Instead of dropping them back on the snow and moving on, you stop to think. You ask yourself why the insects died when you picked them up. After thinking about this for a while, you come up with several possibilities.

First, you think, "Maybe they did not actually die." You remember hearing about some insects that "play dead" as a way of avoiding being eaten by predators. This should be easy to test, you think; all you have to do is take some with you and see if they revive. But they look dead as dead can be, so you think some more. They were out in the winter, crawling on the snow, so they are probably acclimated to cold temperatures. Perhaps your hand warmed them too quickly and the rapid rise in temperature killed them. How would you test this?

Maybe you could try picking up a few more, but this time keeping your mittens on so your hands do not warm them. You do this, and they survive! This supports your hypothesis, and suggests that these winter insects could not withstand the relatively high temperature of your hand. But now that you have started thinking about this, you are not satisfied. You keep thinking of more possibilities. Suppose it was not the *warmth* of your hand that was the culprit. You have been putting lotion on your hands, and it is possible that the lotion is toxic to the bugs.

In fact, this hypothesis also predicts that the insects should survive if you pick them up with mittens, so a clever test will be needed. You are well on your way!

This is the essence of scientific thinking. Observations lead to questions, and questions lead to hypotheses. Remember always that hypotheses are formulated for one reason: to be rejected.

9

This means that, as you think of hypotheses, also think about how you would test or reject them. We discuss hypotheses in more detail in Chapter Six.

CHAPTER THREE

Kinds of Data and Scales of Measurement

O nce you have selected an appropriate (and interesting!) research question, you need to start thinking about how to answer it. In particular, you will want to consider the type of information that will be most useful for discriminating among your hypotheses.

Let us start by asking the very simple question, "How many students are in my biology laboratory section?" Obviously, all we must do to answer this question is make a simple head count. If we had asked a question of a comparative nature, such as whether there are more men than women in the class, the data collection would be slightly more complex.

In this case the population under study (the class) would have to be categorized into two groups (men and women), each group would have to be counted, and the totals compared. You will note the term "population" was used to describe the subjects of the latter study. Most things come in populations, or groups of similar kinds of things. You could use the term when discussing cars, birds, plants, numbers or people. No matter what population of items you study, the kind of question you ask determines the kind of data you collect.

Although it was perfectly obvious in the two sample questions about the students in your class, it must be stressed that, for any test to measure the property being considered, the data collected must be relevant to the question. There are different kinds of data and different scales of measurement.

Kinds of Data and Scales of Measurement

When we identify or estimate various **parameters (aspects)** of a **population** (a group of similar kinds of things), we may use one of four different scales of measurement. These scales of measurement differ in their power to provide us with information and in the degree to which they supply us with data susceptible to statistical analysis.

The first two scales of measurement, the **nominal scale** and the **ordinal** (or **ranking**) **scale**, are appropriate for **discrete data**. In discrete data, each item is a separate, whole unit. Cats and dogs are discrete units because there is no possibility of a cat-and-a-half falling between cat and dog. The head count of members of your class was a measurement of discrete data.

When discrete data are measured on a nominal scale, they are merely collected and grouped by arbitrary names, numbers, or symbols. A population of cars could be divided into groups named Honda, Chevy, Ford, etc. People could be grouped by sex into the categories male and female. The classifications of nominal data are equally weighted. The category names (such as Honda or Chevy) have no particular relationship to each other except that they describe the same population. Even when numbers are used to describe categories of nominal, discrete data, the numbers are arbitrary—rather like the numerals on the shirts of football players. (Football numerals could be considered equally weighted because the player with the highest number is not necessarily the heaviest, fastest, or best; the numbers are merely identification.) **Nominal measurement is merely a sorting of a population into named groups**; it is the weakest form of measurement. Nominal measurement can be used to answer questions of how many and how frequent.

You might identify the butterflies in your backyard by their species name and count how many you find of each type. This would be a measurement of discrete data on a nominal scale. The data would only tell how many you found in each category and which species occurred most frequently. More complex questions require a more powerful scale of measurement.

Some nominal, discrete data can be organized along an ordinal or ranking scale. **When a population is measured on an ordinal scale, the named categories are organized in terms of some relationship they have to each other.** Order can be imposed on the nominal data about cars by ranking the categories by weight, for example. VWs might be the lightest category, followed by Honda, Chevy, Buick, etc. Note that the items, although organized into a relative order, are still discrete. There will be no Chevy-

and-a-half on the scale. Nominal categories such as military ranks (sergeant, major, etc.) are also examples of categories that may be ordered on a scale of relative value or importance. When things are measured on an ordinal scale, the resulting data, besides yielding information about how many and how frequent, can also give you information about a central tendency called the **median**. (The median, which will be discussed more fully in Chapter Four, is the middlemost score or rank in a series of ordinal measurements.)

If you do an experiment in which you test a hypothesis about a correlation between the discrete data you have collected on an ordinal scale and some other phenomenon (such as a test to see whether there are more cars in the heavier categories), ranking statistical tests apply. (The Spearman's Rank Correlation test in this handbook is an example.)

The first two scales of measurement we have discussed, the nominal scale (a naming of categories) and the ordinal scale (named categories are ranked in some order), are both suitable for discrete data in which each item is a separate, whole unit. More information can be obtained by measuring on an **interval** or a **ratio scale**. Both of these scales of measurement are for **continuous data**. Continuous data are points taken along a scale that, at least theoretically, could be subdivided.

If you were noting times of a particular phenomenon and your readings were 7:00, 7:30, 8:00, 8:30 . . ., there always would be the possibility of additional data points such as 7:31, 7:45, etc. Unlike discrete data, which involve whole units such as cats or cars, **continuous data are taken from some continuum** such as time, weight, or temperature. Additional data points between those observed and recorded are always theoretically possible. As we have said above, there are two scales of measurement for continuous data, the interval scale and the ratio scale. Unlike the ranking scale, which merely gave an order to named categories, **the interval and the ratio scale indicate the distance between the items.** The intervals, or distances between the categories, are purely arbitrary, and the ratio between any two intervals is independent of the unit of measurement and of the zero point. (We will illustrate this in a minute.) The only difference between the interval scale and the ratio scale is that, in the interval scale, there is no actual, "real world," zero point.

The most common example of an interval scale of measurement is the measurement of temperature. Temperature is customarily measured on either one of two different interval scales called Celsius and Fahrenheit. Because there is no such thing as a complete absence of temperature, there is no "real world" zero point. In each of these interval scales an arbitrary zero has been established (based on the temperature at which water freezes in the case of the Celsius scale).

To illustrate, let us assume that these temperature readings represent some data collected in both Celsius and Fahrenheit. Notice that these two interval scales and their zero points are arbitrary, but the intervals, or gaps, between readings are proportional.

| CELSIUS | 0 | 10 | 30 | 100 |
| FAHRENHEIT | 32 | 50 | 86 | 212 |

If you do not see that the data are proportional, note that the size of the gap between $0°$ and $10°$ (10) and between $10°$ and $30°$ (20) is the same ratio (10:20) as that of the corresponding gaps in the Fahrenheit readings (18:36). In other words, the differences between temperature readings on one scale are proportional to the equivalent differences on the other scale, but the numbers themselves are arbitrary.

It is theoretically possible to create an interval scale using non-numerical categories since the unit of measurement is arbitrary. But, before any meaningful statistical manipulation could be performed, it would be necessary to establish that the intervals or distances between the categories were equivalent. In fact, it would be necessary to translate them into numbers before any arithmetic operations could be performed.

Data measured on an interval scale are susceptible to more statistical treatment than are nominal or ordinal data. Because the interval scale is truly quantitative, many statistical parameters (aspects) may be estimated. Interval scale measurements can yield information about means, standard deviations, and

14

correlations and can be evaluated by all the common parametric statistical tests.

The other powerful, quantitative, and statistically useful scale of measurement for continuous data is the ratio scale. **Measurements on a ratio scale differ from those on an interval scale only in the respect that they always have a true zero point**. Once again the actual units of measurement (inches, millimeters, etc.) are arbitrary. In the following example of continuous data measured on a ratio scale, note that the measurements in inches and in millimeters share a common zero point. And beyond that point, the units of measurement are arbitrary and, again, proportional.

INCHES	0	1	10	20
MILLIMETERS	0	25.4	254	508

Bear in mind that the kind of data you collect in the experiments you design will dictate the way the data can be analyzed statistically. **It is important to consider from the start what you will measure and how you will treat your data.** If you are merely grouping cars by kind to see how many of each kind go by the window, you are collecting nominal, discrete data. If you are interested in whether more people drive heavy cars, you may have ranked your nominal categories on an ordinal scale by relative weight. If you are stopping each passing car and driving it up on a scale to record actual weights, you are collecting continuous data which will be measured on a ratio scale.

If you find later that you need discrete data, but you are still interested in using the weights you have recorded, you may wish to distribute your data into some arbitrary, but discrete units of measurement such as 0 to <100 lbs., 100 to <200 lbs., 200 to <300 lbs., etc. It would now be converted into a discrete, ordinal scale.

Understanding what kind of data you need to answer a particular question, and being able to show statistically that your results were not merely the result of chance, involve advanced planning and an understanding of the concepts presented in this chapter.

CHAPTER FOUR

Description of Data:
Central Tendencies

Y ou will need a way of interpreting whatever kind of
data you gather. If you have taken a random sample
of a population, estimating one or more parameters
(aspects), you now have a series of readings. Suppose that
you have measured the height of each tree on a previously
unexplored island. Actual tree heights are continuous data,
measured on a ratio scale, but to simplify your experiment,
you have limited your measurements to discrete units of
one meter each.

One of the most useful and illuminating first steps in
organizing data is to plot it on a graph so that it can be
visualized more easily. For many types of experiments,
data can be plotted on a **frequency histogram**.

Traditionally, the vertical axis (called the **ordinate**)
stands for the frequency of occurrence of a particular
measurement. The horizontal axis (called the **abscissa**)
always represents your chosen units of measurement. The
lowest measurement is always plotted at the extreme left;
the measurements get progressively higher as you proceed
to the right. Figure 4.1 is an example of a frequency
histogram for recording tree heights measured in one meter
units.

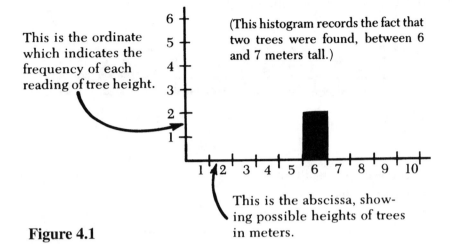

This is the ordinate which indicates the frequency of each reading of tree height.

(This histogram records the fact that two trees were found, between 6 and 7 meters tall.)

This is the abscissa, showing possible heights of trees in meters.

Figure 4.1

Once you have plotted your data, it may be obvious that it is distributed with some particular pattern of frequency. It is possible, but unlikely, that your data will be uniformly distributed. A **uniform distribution** of data would look like this:

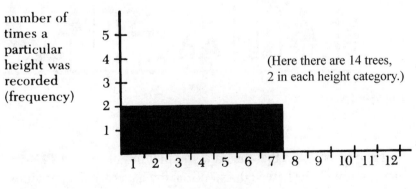

number of times a particular height was recorded (frequency)

(Here there are 14 trees, 2 in each height category.)

Figure 4.2

heights of trees in meters

It is important to understand that if the data looked like Figure 4.2, the trees did not look like this:

Figure 4.3   This is a measure of height in meters.

If the woods had looked as we have pictured it in Figure 4.3, the data points would have fallen into a single, tall column over the 2 meter mark on the abscissa.

The uniform distribution of tree heights plotted in the histogram (Figure 4.2) might have come from a woods that looked like Figure 4.4 (there are an equal number of trees in each category).

This is a measure of height in meters.

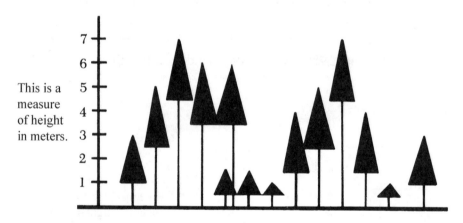

Figure 4.4

It is also possible that the data you plot might show a **skewed** distribution. In a skewed distribution, the data reveal an obvious central tendency, but they are not evenly distributed on either side of the high point. Data can be skewed to the right, in what is called a positive skew, or to the left, in a negative skew. In our example of skewed data (Figure 4.5), we see a negative skew; the low readings are more extreme than the high ones.

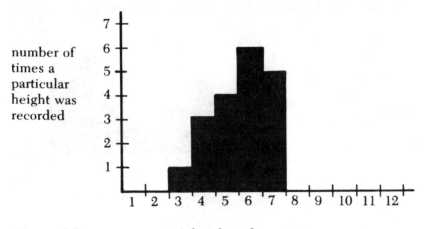

Figure 4.5   heights of trees in meters

number of times a particular height was recorded

The trees represented on Figure 4.5 might have looked like those in Figure 4.6—there are many trees in the taller categories.

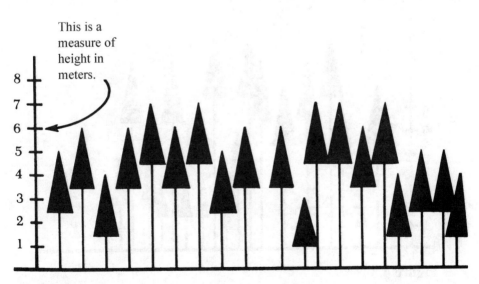

This is a measure of height in meters.

Figure 4.6

Another kind of distribution which your data might display when plotted on a frequency histogram is called a **bimodal distribution.** In a bimodal distribution the data seem to fall into two groups. In Figure 4.7 we have plotted another histogram of tree heights—this time with a bimodal distribution.

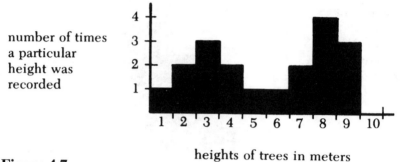

number of times
a particular
height was
recorded

heights of trees in meters

Figure 4.7

Often when your data falls into a bimodal distribution, there is reason to suspect that the population you have been studying is really two populations. The woods represented in the bimodal distribution in Figure 4.7 might have looked like this:

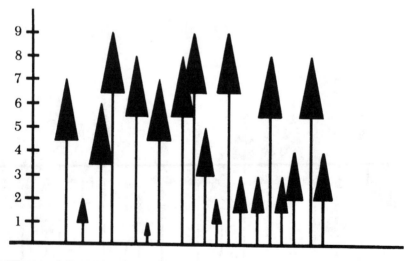

Figure 4.8

20

Or it might have looked like this:

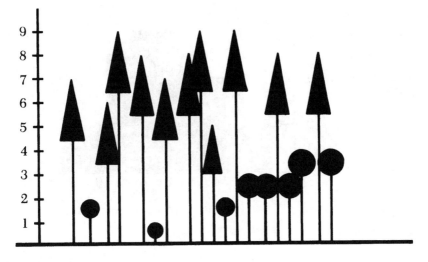

Figure 4.9

Perhaps you were treating two different species as if they were one. Even when you are clearly dealing with only one species, such as *Homo sapiens* (man), you might plot your readings of adult body weights and find that they fall into a bimodal distribution. In this case your readings would indicate two central tendencies, one for men and one for women.

There are other ways data can be distributed: random distributions, log normal distributions, etc.; but the most frequent kind of distribution found for a sample of a single population is called a **normal distribution**. It is equally spread out on either side of a central high point with a characteristic "bell" shape. Figure 4.10 is an example of a normal distribution of our trees.

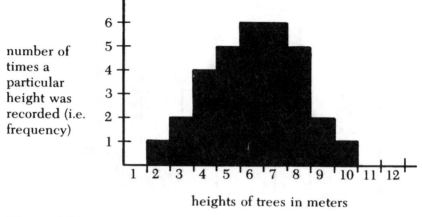

heights of trees in meters

Figure 4.10

Most data, even if somewhat skewed, are spread out on either side of a central high point. When you analyze your data, you are often interested in reporting what usually happens—in other words, the central tendency. There are three common measurements of central tendency—the mean, the median and the mode.

The **mean**, or numerical average (often written as $\overline{X}$), is the simplest measurement to make. All you have to do is add up your various measurements and divide this total by the number of measurements taken. To calculate the average, or mean height of the trees on your study island, you would add up all the different heights and divide this sum by the number of trees in the sample measured. The mean is useful in many statistical tests, but, reported alone, may not be very meaningful. If your plotted data were very skewed or bimodal, the mean might be deceptive. A mean is always strongly affected by any extreme readings.

Another measurement of central tendency, one that is less susceptible to distortion by an extreme reading, is the **median**. Before you can calculate the median, you must organize your readings in a progressive sequence. If you were going to calculate your median tree height, you would put all your measurements in order with the shortest first and the tallest last. The median of the distribution of data is the middlemost value. When you have an even number of readings, the median falls between the central figures. You can calculate an exact median by averaging the two central data points (adding them together and dividing by two).

Once you have your data organized sequentially, you can easily find the **mode**. The mode is the most frequently occurring value. There can be more than one mode to a distribution as we have seen in the example of the bimodal distribution.

If your data fit a *perfect*, normal distribution, the mode, the median, and the mean would be the same figure, *by definition*. However, in many real distributions of data, that are called "normal" (because for statistical purposes they do not differ significantly from the theoretical normal), these three measures of central tendency need not be the same, and often they are not. If your measurements were the following: 2, 3, 4, 4, 5, 6, 8, 10, and 12, the following would be your measurements of central tendency:

the **mean** would be 6 (the sum of the values: 54, divided by the number of values: 9);

the **median** would be 5 (the middlemost value); and

the **mode** would be 4 (the most frequently occurring value).

CHAPTER FIVE

Description of Data: Dispersion

We have indicated some of the ways to measure a central tendency in your data. Often it is useful to know what is happening outside the central tendency—to have some measurement of the dispersion or spread of the data about the mean ($\overline{X}$). Dispersion can be measured in terms of the **range** of the data; this is merely the "distance" between the lowest and the highest readings.

Let us assume that you have measured the height (in meters) of seven trees, selected at random, from each of four different islands. Your data, when organized from the lowest to the highest reading might look like this:

Island I	8	8	9	10	11	12	12
Island II	5	6	8	10	12	14	15
Island III	1	2	5	10	15	18	19
Island IV	8	10	10	10	10	10	12

Figure 5.1

Note that each island had the same average tree height. If the means were the only data you reported, the islands might seem remarkably similar. In fact, there was a considerable variation between them. Island I and IV had a range of 4 meters, and Island III had a range of 18. In this instance, calculating the range of the data might be quite important. Here is a display of the data from our four islands reporting the means and the ranges of tree height in meters.

Island I	$\bar{X}$ = 10 M	range = 4 M
Island II	$\bar{X}$ = 10 M	range = 10 M
Island III	$\bar{X}$ = 10 M	range = 18 M
Island IV	$\bar{X}$ = 10 M	range = 4 M

Figure 5.2

However, reporting only the mean and the range does not tell anything about how clustered the readings were. The bulk of your data might not deviate much from the mean. It is nice to know how many readings were close to the mean and how many were widespread. The other measurement of dispersion, or spread in data, is called the **variance**. It is a statistically more useful measurement which conveys more information than the average or the range alone. To understand the concept of variance, it is necessary to know more of the mathematical properties of the normal distribution. (The formula for calculating the variance will be given on page 31.)

Although your data will never exactly coincide with the perfect, normal distribution, they may be fairly close. It will usually be appropriate, therefore, to describe them in the terms we will give for a normal distribution.

The data in Figure 5.3 are normally distributed. Again the frequency is plotted on the ordinate, and the unit of measurement (whatever it may be) is plotted on the abscissa.

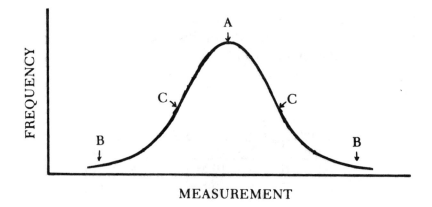

Figure 5.3

You will note that the step-like characteristics of the histogram have been removed because, in this plotting of data, all the theoretical points between each step are represented, smoothing the steps into a line. (Data are often displayed with a curved line even when not all the intermediate steps have been recorded.) On this curve, often described as a bell-shaped curve, the highest point (A) is the mean ($\overline{X}$), the median, and the mode. The two tails of the curve (B) are not drawn touching the baseline. The two lines (tail B and the abscissa) are said to be **asymptotic** to each other. They are presumed to recede indefinitely, coming closer and closer without ever touching.

There is another mathematically and theoretically important feature to the normal curve, the **points of inflection**. The points of inflection (C) are points at which the curve changes from convex to concave. For the purpose of demonstrating the variance, theoretical lines can be drawn to the abscissa from the two points of inflection and from the mean point. The distance between the mean point on the baseline and the point of inflection, on the baseline, is called one **standard deviation**. It is a sort of average of the deviation from the mean. The symbol for the standard deviation is "s."

26

Here is another normal curve with the standard deviations drawn.

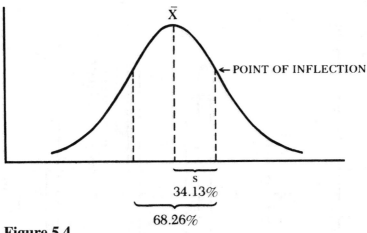

Figure 5.4

Of the data you collect from a normally distributed population, 34.13% *by definition*, falls within one standard deviation from the mean. One standard deviation on either side of the mean includes 68.26% of your data.

It is very helpful to report the range and the standard deviation of your data. A particularly clear way to summarize data taken from more than one population, without having to plot all the different readings, looks like Figure 5.5. Here we have taken the data from Figure 5.1 for the trees on the four islands and shown the means, the ranges, and the standard deviations. (Note: there is no abscissa—the vertical scale gives tree heights in units of one meter.)

Figure 5.5 clearly displays the fact that the average height of each population is the same. In addition, the extreme variation in the range of the data is obvious. The boxes which show the standard deviation on either side of the mean help the reader understand how clustered or dispersed the data were (68.26% of the readings fell inside the boxes).

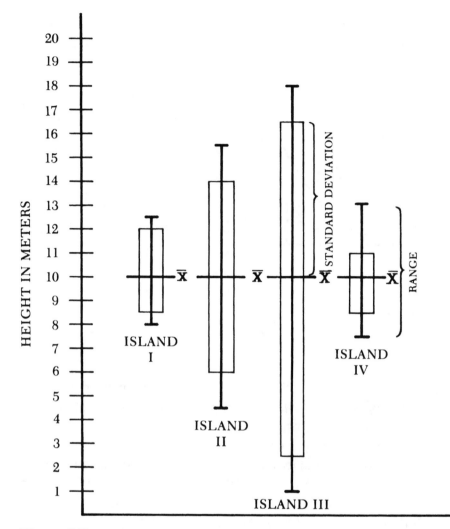

Figure 5.5

Sometimes it is useful to talk about "two standard deviations from the mean" or even three. The standard deviation is an arbitrary, fixed length on the abscissa. If you measure out two standard deviations from the mean in each direction, you have accounted for 95.44% of the data. Three standard deviations would include 99.74% of the data. Any reading which fell outside three standard deviations from the mean would be an extreme one indeed. However, there always is the probability of such a reading.

Here is the normal curve again, with three standard deviations drawn on either side of the mean.

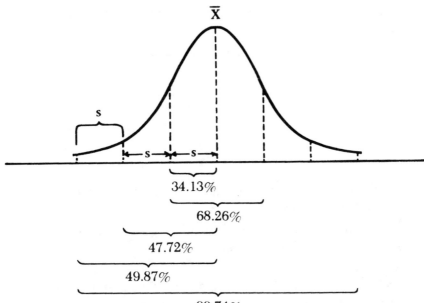

Figure 5.6

Because the standard deviation is a particularly useful and important aspect of your data, it is helpful to have an easy means to calculate it. It is not done by dropping lines from the inflection points on the curve of your data because the inflection point is a theoretical point, and your data will not fit the true, hypothetical normal curve.

There is a mathematical formula for calculating "s" (the standard deviation). The formula is appropriate for any data that are fairly close to being normally distributed. (If you are doubtful about your data, even after you have tried plotting them on a histogram, mathematical tests can evaluate the distribution of your readings.) Although the formula looks complicated at first glance, you can use it whether or not you understand it. The data are merely inserted according to the directions. (In this handbook, none of the mathematical calculations are difficult ones and previous mathematical training, although useful for the ultimate understanding of such formulas, is not required.) Some calculators are programmed to calculate the standard deviation for you.

This is the formula for calculating the standard deviation of your data.

$$s^2 = \sqrt{\dfrac{\Sigma x^2 - \dfrac{(\Sigma x)^2}{n}}{n - 1}}$$

s = the standard deviation (this is the figure you will calculate, you do not plug it in at the start)
Σ = the mathematical symbol for sum
x = any one of your data points
x^2 = x times x
Σx^2 = square each of your data points, then sum these squares
$(\Sigma x)^2$ = the square of the sum of all your data points (add the data points and then square the sum)
n = the number of the data points you have collected
n - 1 = one less than the number of points collected

In other words,
1) square the values for each of your data points and then sum these squares, Σx^2;
2) sum the values for each of your data points and then square this sum $(\Sigma x)^2$;
3) divide the value obtained in step 2 by the number of data points, $\dfrac{(\Sigma x)^2}{n}$
4) subtract the value obtained in step 3 from that obtained in step 1;
5) divide the result of step 4 by the number of data points minus 1; and
6) take the square root of the figure you have calculated. This is the standard deviation of your data.

If you now look at the data on tree heights in Figure 5.5, you will see that in populations with a high degree of similarity, the standard deviation is small. In highly dissimilar populations, the standard deviation is larger. This is why it is such a valuable measure of the dispersion of data.

As we have indicated earlier, the variance is one other frequently used measurement for describing the dispersion of data. The **variance** is defined as the square of the standard deviation, or s^2. The formula for calculating the variance is the same as the calculation of the standard deviation except that there is no need to take the square root.

$$s^2 = \frac{\Sigma x^2 - \dfrac{(\Sigma x)^2}{n}}{n - 1}$$

Because the square root is not taken, the variance is usually too large to plot in the kind of data display we used for the four tree populations in Figure 5.5. The variance is a figure you may need to calculate for certain statistical tests. Some tests use the variance and some use the standard deviation.

In the next few chapters we will introduce the idea of the null hypothesis and then proceed to the analysis of data to test the null hypothesis.

CHAPTER SIX
The Experimental Plan

One of the most difficult tasks for the beginner is the selection of a research topic. It is absolutely essential that the initial question chosen can be answered through simple observation or experimental manipulation—using relatively unsophisticated techniques—in a reasonable amount of time. If you are in a research course, a general menu of available species and equipment will probably be supplied. If you are on your own, you might wish to examine some books about research problems and laboratory techniques (a few are mentioned in the bibliography at the back of this handbook). Or, you can simply find a topic by gradually narrowing the field of possibilities until you hit upon something feasible.

You might ask yourself whether you are interested in plants or animals. If animals, large or small? Unless you live on a farm, in a forest, or at the zoo, it is wise to select a relatively small species that is abundant and easy to find. Again, your geographical situation should help you choose between terrestrial or aquatic animals, etc. Be guided by the practical aspects—what is actually available to you in nature, in somebody's laboratory, at the pet shop, or from a biological supply house.

Once you have focused on a handy species, say some local species of ant, read about it, talk to people, and see, for example, if you are interested in its anatomy, physiology, or possibly some aspect of its behavior. Think about how you would study this species. Would you work outdoors? In your apartment? Would you need to construct some housing such as glass-covered trays? These decisions will depend on the specific question you decide to ask. Eventually, you must focus your attention on a very small aspect of the ant's repertoire, and ask a question you can test, such as: How long does the scent marking on an ant trail last?

In most biological experiments there are numerous sources of error. For example, let us suppose that you are studying the effect

32

of light on the amount of CO_2 taken up from an aquatic medium by an underwater pond plant. You may be using, or misusing, an electronic CO_2 probe to take readings of the amount of CO_2 in the water. You may also be measuring the light intensity with a light meter which may or may not be working properly. Your experimental plant may be in poor health without your realizing it. There may also be some problem with the logic of your experimental design, or you may be subjecting your data to an inappropriate statistical test. If for some reason, your results are unintelligible, you will not know whether the fault lay in your measurement (either a human or mechanical failure) of the two parameters (light and CO_2), in the biological system being examined (the plant), or in your experimental plan of attack.

Therefore, we will concentrate here on the very basic elements of experimental planning. Chapter Nine, which outlines a search strategy for getting information out of a library, will help you find information relevant to your own particular topic of interest.

I. The first step in planning an experiment is choosing an answerable question.

Once an area of interest has been identified (I'm interested in the people passing by the door of this lab), you must focus on a specific question (Are there more men than women passing the door?). Your hypothesis may be that there are more men than women; often an experimenter has something in mind that he expects to show. The trick in good experimental design is to capture the phenomenon in a logical box.

You will recall from the first chapter that all progress is made by the rejection of hypotheses. There is no such thing as proof. **All the possible alternative explanations for a phenomenon or all experimental outcomes must be listed and an attempt made to eliminate each possibility.** (In our example in the first chapter, all explanations except the too-little-sunshine hypothesis were rejected, permitting us to infer that the lack of adequate sunshine was the factor causing the death of the ferns.)

Because the logical experimental framework is based on the rejection of hypotheses, experiments usually begin with the statement of what is called a null hypothesis (abbreviated H_0). **The null hypothesis is a hypothesis of no difference.** If we are interested in whether more men than women are passing the laboratory door, our null hypothesis is,

H_0: There is no difference between the number of men and the number of women passing the door. (For statistical purposes the numbers are the same.)

Naturally, you will collect data on the number of people of each sex passing the door. You may count more men than women and consider the experiment over. The absolute values you have counted may be different, but this difference may be so slight that it is *statistically meaningless*. Although you may have counted more men, chance alone could have caused this difference, and it may not tell you anything important about the population under study. As a scientist, you should test your data to see if you have a statistically significant reason to reject your null hypothesis.

If you fail to reject your null hypothesis, your experiment is over. You have shown that, even though your absolute values may have differed, there is no meaningful difference between the number of men and the number of women passing the door. However, if your statistical test has shown a significant difference between the two absolute values, you may reject your null hypothesis. This is an important step forward.

If you had only been interested in showing that there *was* a difference, your first (and, in this case only) alternative hypothesis would have read:

H_a (alternative hypothesis): There is a difference between the number of men and the number of women passing the door.

Normally, as you plan your experiment, you reduce your plan to a symbolic shorthand format (because it will then be easier to substitute the actual data collected for parts of the formula). The logical framework for your experiment might look like this:

H_0: # men = # women   (# means "number of")

H_a: # men $\neq$ # women ($\neq$ means "not equal to")

However, it is likely that you would want all the possible alternative answers as part of your logical framework—one alternative (H_1) stating that there were more men and the other (H_2) that there were more women. Your experimental plan would have looked like this when reduced to its logical format:

Question: Are there more men than women passing the door?

H_0: # men = # women

H_1: # men > # women

H_2: # men < # women

(The symbol > means "greater than" and the symbol < means "less than.")

If the outcome of your statistical test permits you to reject your null hypothesis, and if you have listed all the possible alternatives, you are now in a position to pick one of the alternatives. (In the above test it was easy to list all the alternatives because there were obviously only two and they were clearly mutually exclusive; but in the fern experiment in Chapter One, there were many possible alternative hypotheses.)

In some instances, further experimentation may be involved before you can distinguish between alternatives. Under most circumstances, if all logical possibilities have been covered, all you need to do to select the correct alternative is to examine your data. In our people-passing-the-door experiment, once we have rejected the null hypothesis, if we have counted more men than women passing the door, we now have a statistically significant reason for concluding that more men than women pass the door. The difference in absolute values, the actual counted numbers, was probably *not* a chance event. You will see later in the book that certain statistical tests such as the Kruskal-Wallis Test, the Friedman Test and the Analysis of Variance permit you to reject your null hypothesis but do not, in themselves, allow you to select the correct alternative hypothesis.

To summarize the first step in planning an experiment, you must begin by establishing your logical framework. Your experimental question must be stated in terms of a null hypothesis and all logical alternative hypotheses must be listed. You should then be able to reduce these verbal statements to the simple, logical shorthand format shown in the example above. Once the logic of an experiment has been determined, you may proceed to the second step.

II. The second step in an experimental design is determining what kind of data will be needed to answer your question.

Here the concepts of discrete and continuous data become relevant. Besides knowing **what kind of data** you will collect, you should know **how much data** to collect. The question of how much data is more important than you might think. Of course, time can be wasted by collecting too much data, but the real danger is that you may not collect enough.

Consider this, for a moment. If you were flipping a coin and, in 10 tosses, you got six heads and four tails, you would have no strong reason to suspect the coin was defective or unfairly weighted. However, if in 1000 tosses, 600 were heads and 400 tails, you would have reason to be suspicious. It was the fact that the 6:4 ratio was the result of a great many tosses that caused the alarm. We will go into this more fully later, but the example is intended to show that a large sample size is more convincing and meaningful in mathematical terms. The tables which accompany the statistical tests in Chapter Eight will be useful in determining appropriate sample sizes for your data.

All the statistical tests are merely ways to examine different kinds of data, collected in different ways, to determine whether you have a statistically significant reason to reject your null hypothesis.

The important part of step two, in which the plan of attack is outlined, is that you know ahead of time how you will test your data. The statistical test to be used may require a certain kind of data (ordinal, perhaps) and certain minimum sample sizes (the minimum amount of data you must collect). The entire experiment should be thought through at the start.

Whether you do this as a means of organizing your thoughts,

or to have a research topic approved by your instructor, or because you are applying for financial aid, it is important to write your research plans in the form of a **research proposal**.

The research proposal should contain a clear description of the areas of interest and of the specific question being asked experimentally. An explicit statement of the null and alternative hypotheses is also useful. A full written proposal should also show a justification of the effort (and perhaps expense) entailed. This usually indicates why the question is important and how it fits into a broader picture of what is already known about the topic. Normally, a full search of the relevant published literature precedes a serious experiment. (Chapter Nine deals with a search strategy for finding such information in a library.)

The proposal should outline the actual plan of experimental attack, mentioning specific equipment and techniques to be used (and referring to the literature if previously published procedures are to be followed). The organization of the controls and variables should be described, as well as the sample size envisioned. For students, an actual day-by-day schedule of events and procedures is particularly useful. If the experiment involves several steps, a verbalized or diagrammed flow chart of decision points is helpful as well. (If A happens, we will do such-and-such, if B, we proceed to....)

Finally, the proposal should mention the statistical treatment that will be applied to the data collected. (Chapter Seven will help with the selection of the right test for your kind and amount of data.) It is only with this type of careful planning that you can avoid collecting meaningless, inconclusive or irrelevant data.

Later, when you are writing the report of your experiment (see Chapter Eleven), you will find that most of the material in your proposal will be useful. The justification may be used in writing the introduction and possibly the conclusion; the description of methods and materials will be practically written; and the results section will follow logically from your statistical plan. The time spent on the research proposal will be well invested, insuring a logical experiment and an easily written report.

We recommend the research proposal format on the following page.

Name_____ Lab Section_____

TITLE:

(For help in writing a descriptive title, see Chapter Eleven.)

QUESTION:

WHY THIS QUESTION:

EXPERIMENTAL LOGIC:

H_0 (null hypothesis):
H_1 (first alternative):
H_2 (second alternative):
(There may be more alternatives.)

METHODS AND MATERIALS:

PROPOSED STATISTICAL TREATMENT:

(Sometimes it is useful to make up hypothetical data of the type you expect to collect and actually try the statistical test. This may reveal how much data would make an adequate sample.)

CHAPTER SEVEN
The Statistical Analysis

In Chapter Eight you will find your arsenal of statistical tests. It is possible that you will require others not given here, but this selection will cover most of the beginning experimental situations. The collection begins with a flow chart (page 50) to help determine the appropriate test. The flow chart guides you through decisions which depend on the kind of statistical question you are asking and kind of data you are collecting. Each test is preceded by precautions which limit the test to certain kinds of data. Each test is followed by a sample experimental situation in which the test is used appropriately.

Learn to use these tests as you would use a reference book— without memorizing the contents. You only need to know how to select the correct test and then plug in your own data.

Because all of the tests are aimed at determining whether you have a statistically significant cause to reject your null hypothesis, they are all asking either one of two general kinds of questions— questions of correlations or questions of differences. On the flow chart, the first decision you must make is: Which kind of questions are you asking your data?

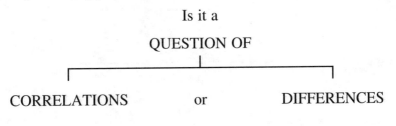

Is it a

QUESTION OF

CORRELATIONS   or   DIFFERENCES

Questions of Correlations

This sort of statistical question is often asked when you have examined two parameters of a population (such as the height and weight of a group of people) and you want to know whether there is a meaningful **correlation** between these factors. And if they are related, is there a positive correlation (taller people are heavier) or a negative correlation (taller people are lighter)? When you are looking for a correlation between two factors, the data are plotted differently from the frequency histogram we have shown you. In this instance, one factor, such as height, is measured on the ordinate and the other factor, weight, on the abscissa. Each measurement you have taken would be recorded as a dot on the graph over the correct weight at the level of the appropriate height.

Data which are positively correlated might look like this:

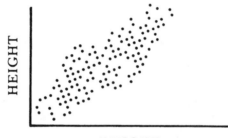

Negatively correlated data might look like this:

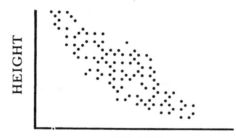

Data which are not correlated would look like this:

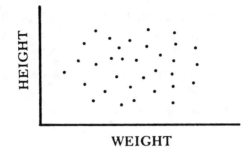

Often the data are nowhere near as obvious. A statistical test is needed to permit you to reject your null hypothesis which, in questions of correlations, would read: There is no correlation between . . . (height and weight, in our example).

When two factors are correlated, the magnitude of one changes with the magnitude of the other, but no cause and effect relationship need exist. Height does not *cause* weight nor weight cause height, although the factors are clearly related.

In some situations, however, there may be a dependent relationship between two variables. The magnitude of one variable may be dependent on the magnitude of the other. This is often true in the case of human blood pressure. A person's blood pressure (the dependent variable) may be a function of his age (the independent variable), although his age would not be a function of his blood pressure. One variable depends, to some extent, on the other. **Regression analysis** is used to study the strength of this kind of relationship between two variables. Regression analysis may also be used to determine the extent to which one variable, such as SAT (Scholastic Aptitude Test) scores, can be used to predict values of another related variable, such as college grade point average.

Questions of Differences

At your first decision point on the flow chart, you needed to know whether you were testing for correlations or differences. If you are testing for differences, you must choose your statistical analysis from one of three categories—differences between distributions of discrete data, differences between means (of two or more samples) and differences between variances. We will consider each of these categories briefly.

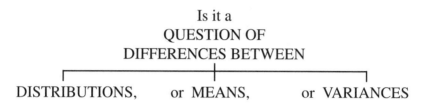

A test for **differences between distributions** of discrete data would tell you whether some data distribution differed either from some theoretical distribution or from some other actual distribution. In the first instance you might be testing the heads:tails ratio from a coin toss (say 55:45) to see if it were statistically different from the theoretically expected ratio of 50:50. Your null hypothesis in this instance would be:

$$H_0: 45 : 55 = 50 : 50$$

If you were looking for differences between two actual distributions of discrete data, you would be testing for what is termed "independence between two samples." You might be comparing data collected on the ratio of men to women living in Town A (5,467 men in Town A : 5,688 women in Town A) to the ratio found in Town B (8,975 men in Town B : 9,230 women in Town B). For testing independence between two samples using this example, your null hypothesis would be:

$$H_0: \text{\# Town A men} : \text{\# Town A women} = \text{\# Town B Men}: \text{\# Town B women}$$

or

$$H_0 : 5,467 : 5,688 = 8,975 : 9,230$$

Tests for **differences between means** ask the data if the mean of one population distribution is significantly different from the mean of another population distribution. This could be asked for two populations which you suspect might be different or, when you have separated one population into a control group and an experimental group, and you need to know whether your experimental manipulation has created a meaningful difference in their means. Let us say that you plotted the heights of a control group of plants (A) and an experimental group (B), which differ only in that they have been treated with a growth hormone. Your null hypothesis is this: H_0: $\overline{X}A = \overline{X}B$. You are asking which of these two distributions represents the results of your experiment.

This?   or   This?

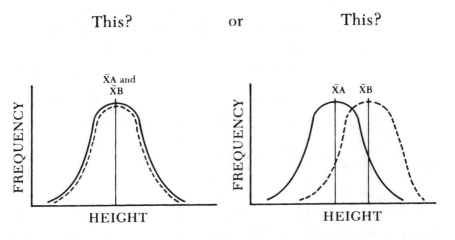

Often tests which compare the means of two populations are making the assumption that the variances (s^2–the square of the standard deviation) are similar. You may need a test to determine whether your data distributions are similar in respect to their variances, either because similar variances are required for some other analysis, or because you are interested in comparing the variances of several distributions for their own sakes. It might be that the frequency distributions of heights of plants in your control group (A) in the plant growth experiment had a mean height of 6 and your experimental group (B) had a mean height of 8, but the variances were 4 and 1 (sA=2, sB= 1). Your distribution curves might look like these:

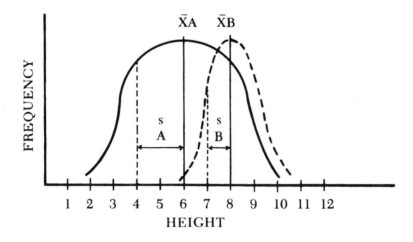

In this instance you would need a statistical test for **differences in variance** to evaluate the following null hypothesis:

$$H_0 : s^2A = s^2B \text{ or } H_0 : 4 = 1$$

The size of the variances often has an effect on your ability to determine statistically whether a significant difference exists between two means. In the plant growth experiment just mentioned, the control had a $\overline{X}$ (mean) of 6 and the experimental group had a $\overline{X}$ of 8. Whether you could reject the null hypothesis that the means were the same would be strongly affected by the size of the variance *even* when they were similar (and not clearly dissimilar as above). The next figure illustrates this point. In the situation on the left, groups A and B have the same variance and means, 6 and 8. On the right, the means are again 6 and 8, the variances are also still equal, but the variances are so much smaller that you might be able to reject your null hypothesis that $\overline{X}A = \overline{X}B$.

44

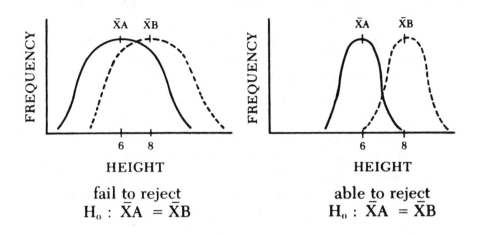

fail to reject
$H_0 : \bar{X}A = \bar{X}B$

able to reject
$H_0 : \bar{X}A = \bar{X}B$

As you study the flow chart for the selection of statistical tests, you will notice that sometimes you need to make a distinction between parametric and non-parametric statistics. In general, **non-parametric** tests can be used to evaluate null hypotheses related to any type of data distribution (normal, skewed, bimodal, etc.); they are not limited to normal distributions.

Parametric statistics are limited to normal distributions. Some statistics limit their application to instances when the following three conditions have been met (although others would relax these requirements somewhat):

1. The data must fall into a statistically normal frequency distribution;
2. The individual observations (or data points) must be independent of each other (you could not use parametric statistics when comparing weights of horse front limbs to hind limbs because the weights would be related, but you could use them to compare front limb weights of race horses to those of draft horses); and
3. The observations must be distributed on the same continuous scale of measurement.

Although it is more difficult to meet these criteria, parametric statistics should be chosen whenever possible because they are often more "powerful" tests. This means that, for a given sample

45

size, they involve less risk that you might fail to reject a null hypothesis that really was false. (They have a greater ability to permit you to reject your null hypothesis.)

Once the flow chart has been followed and the most suitable statistical treatment selected, a scientist normally makes another decision. The scientist must decide how much risk to take that he or she may come to the wrong conclusion about the null hypothesis.

When you are using a statistical test to evaluate a null hypothesis, two types of errors can be made; they are traditionally called the Type I and Type II errors. **A Type I error is the rejection of a true H_0. A Type II error is the failure to reject a false H_0.** For example, you might flip a *fair* coin and get ten heads in a row. Statistically this is unlikely, so your H_0: 10:0 = 50:50 would be rejected and you would conclude from that that the coin was unfair. This is a reasonable mistake, based on a small sample of particularly unusual data. It could happen and it would be a Type I error. You might have thrown four heads and six tails—a much more common occurrence, and failed to reject the null hypothesis that 4:6 = 50:50. In this instance you would conclude that the coin was fair. However, it could happen that a much larger sample size, giving results of 400:600, would have caused you to reject the null. If the coin really was unfair but you had concluded it was fair on the basis of your toss of 4:6, you would have made a Type II error by failing to reject a false null. You can see by these examples that sample size can make a big difference. If you make a Type I error, you jeopardize your integrity as a scientist. A Type II error can also be serious, especially in claims such as: smoking has no effect on human health.

You can never be 100% certain that you have made no mistake, but you do have some control over how certain you will be. In most statistical tests you will probably choose to set your "alpha level" at .05. **Alpha, or α, is defined as the risk of making a Type I error.** When alpha is set at .05, the chances are 1 in 20 that you will accidentally reject a true null. These may seem like pretty favorable odds, but in some testing situations, a 5% chance of failure may be too risky. You have the option of setting any alpha level you wish before you calculate your test. However, a lower alpha level will increase the risk that you will make a Type II error and fail to reject a null hypothesis that really is false. The normal compromise between the risk of Type I and Type II errors is the .05 alpha level.

To illustrate this concept, let us collect data from a *fair* coin. If one test were made up of 100 tosses, and the test was repeated 10,000 times, you would get a frequency distribution of results looking more or less like the one which follows. Each test result ratio (45:65, 42:68, 20:70, 80:20, etc.) would be plotted on a frequency distribution with a center point of 50:50 and the two extremes of 100:0 and 0:100 like this:

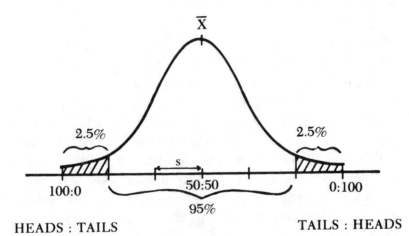

In 95% of the tests, the results will fall inside the white part of the curve (within two standard deviations from the mean); and in 5% of the tests they will fall into the shaded areas (outside two standard deviations from the mean).

Now let us suppose that you performed this 100-flip test only once, using a different coin, and that you were testing the null hypothesis that there was no meaningful difference between the results of your flipping and the theoretical ratio of 50:50. With an alpha level of .05, it *could* happen that you would flip 99 heads to 1 tail, reject your null hypothesis, and conclude that the coin

47

was unfair when it actually was a fair coin. (That is the Type I error.) However, the odds are greater than 95 to 5 that such an extreme reading would *not* be the result of chance alone and that the coin really was unfair. This is the risk you take when using a statistical test to evaluate whether you can reject your null hypothesis.

The best way to avoid making a mistake, particularly a Type II error of failing to reject a null that really is false, is to have an adequate sample size. All of the statistical tests are highly sensitive to sample size. If you are unable to reject your null hypothesis at the α level of .05, you cannot "fudge" your results by changing the α to .09, but you always have the recourse of collecting more data. If the null hypothesis really is false, the increase in sample size will eventually permit you to reject it.

Conclusions

Once you have completed your study and analyzed your results for their statistical significance, it will be up to you to draw your conclusions. The results of a statistical test aren't magical, although you should be hesitant to draw a conclusion about differences between sets of data when no statistical evidence exists for this conclusion. There may be times when statistically significant results may be biologically meaningless—particularly in the case of a poor experimental design. You, as a scientist with your own powers of reasoning, are the most important factor in any study.

The next chapter contains the flow chart to aid in the choice of statistical tests. See the various books on statistics in the bibliography of this handbook for further information and additional tests.

CHAPTER EIGHT

Statistical Tests

This chapter opens with a flow chart for helping you choose the proper statistical test. Chapter Seven defined and explained the steps in the selection process. Is it a question of correlations or differences? If differences, are you testing for differences between distributions, means, or variances? If you are testing for differences between distributions, you must select one of the χ^2 (Chi Square) tests on the basis of the kind of experimental question you have asked. If you are testing for differences between means, you should select a test on the basis of whether you have two, or more than two, samples and whether or not you are in a position to use parametric statistics. All of the tests mentioned on the chart are illustrated in this chapter.

The chapter contains the following statistical tests:

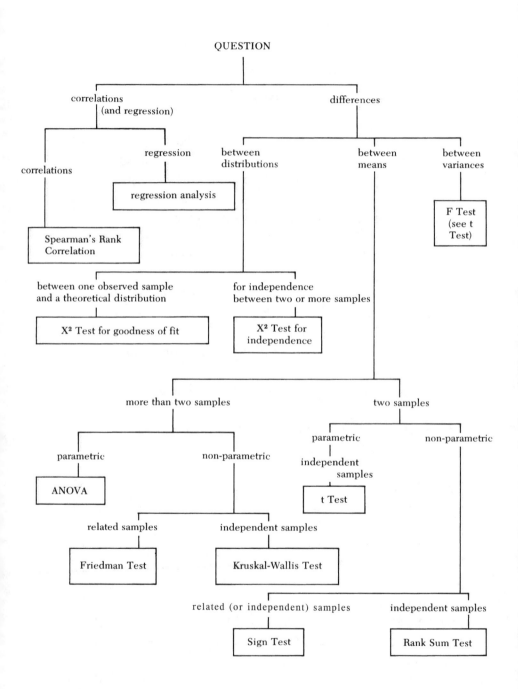

ANOVA — Analysis of Variance

Purpose:

The Analysis of Variance is a parametric statistical test of differences between means of more than two samples. It is designed to test whether the means of the various test samples vary further from the population mean than would be expected when compared with the fluctuation among observations within experimental samples.

Warnings and precautions:

1. The individual observations must be independent of each other.
2. The observations must be from a continuous scale of measurement.
3. The observations must be normally distributed.
4. The variances of the samples must be homogeneous. You may use a "quick and dirty" check for homogeneity by calculating the s^2 (variance—some calculators are programmed to do this step for you) for each sample. Then take the largest s^2 and divide it by the smallest. Proceed with the F test.

Null hypothesis:

The hypothesis to be tested is always:
$$H_0 : \overline{X}A = \overline{X}B = \overline{X}C = \overline{X}D\ldots\text{etc.}$$
If you reject the null hypothesis, you may conclude that the means are unequal when compared with the normal variation within the samples. You have not shown *which* treatment or sample is causing this to be true, however. A least-significant-difference test (LSD), found in advanced texts, will allow you to determine where the significant differences are.

Notation:

The test statistic to be computed is an F ratio. This ratio can be verbally expressed this way:

F = the average of the squared differences between the means of the samples and the grand mean (or mean of all the

51

observations) divided by the variance of all the observations

k = the number of test situations or treatments

n = the number of observations or scores in each treatment

N = total number of observations or scores

x = a data point or score

x^2 = x times x

SS_b = the sum of the squares between groups or treatments (don't worry about what this means)

V_b = the variance between groups or treatments

SS_w = the sum of the squares within groups or treatments (again, don't worry about what this means)

V_w = the variance within groups or treatments

SS_T = total sum of the squares

Σ = the sum

The test statistic, or F ratio we explained verbally, looks like this:

$$F = \frac{V_b}{V_w}$$

To find the values for the numerator and denominator of this formula, we first calculate SS_T and SS_b. SS_w is then found by subtracting SS_b from SS_T. You then divide SS_b and SS_w by the appropriate degrees of freedom to get V_b and V_w.

Procedure:

Step 1. Add up x's of all the treatments to obtain Σx, or the grand total.

Step 2. Square each of the x's for all treatments and then sum these squares to obtain Σx^2. (Note: on some calculators, Steps 1 and 2 can be done at the same time.)

Step 3. Compute the sums and means of the observations in each sample or treatment. In other words, add each of the x's for each treatment and divide by n (the number of observations in that treatment).

Step 4. Compute SS_T by the following formula:

$$SS_T = \Sigma x^2 - \frac{(\Sigma x)^2}{N}$$

where $(\Sigma x)^2$ is the square of the grand total calculated in Step 1.

Step 5. Compute SS_b by the following formula:

$$SS_b = \left[\frac{(\Sigma x_A)^2}{n_A} + \frac{(\Sigma x_B)^2}{n_B} + \frac{(\Sigma x_C)^2}{n_C} \text{ etc.} \right] - \frac{(\Sigma x)^2}{N}$$

This means square the sum of each treatment. (A, B, C etc. stand for the different treatments.) Divide by the number of observations in that treatment, and then sum for each treatment. When this is done, subtract the figure $\frac{(\Sigma x)^2}{N}$ which you calculated in Step 4.

Step 6. Compute SS_w. $SS_w = SS_T - SS_b$

Step 7. Divide SS_b by its degrees of freedom which is equal to $k - 1$, or one less than the number of treatments.

$$V_b = \frac{SS_b}{k - 1}$$

Step 8. Divide SS_w by its degrees of freedom, which is equal to $N - k$, or the total number of observations less the number of treatments.

$$V_w = \frac{SS_w}{N - k}$$

You now have calculated the V_b and V_w values, which are the numerator and the denominator of the F ratio. Calculate F and turn to the F table on page 57. To compare your calculated F with the table, you must consider the degrees of freedom in both the numerator and denominator of the F ratio. The degrees of freedom for the numerator (V_b) is $k - 1$, or one less than the number of treatments. The degrees of freedom for the denominator (V_w) is $N - k$, or the total number of observations less the number of treatments. You will note from the table that the degrees of freedom for the "greater variance" (usually the numerator) is read across the top. The degrees of freedom for the lesser variance, usually the denominator, will be found in the left-hand column of the table. It may be necessary to interpolate between rows to obtain an exact value. If the F statistic you have calculated is *greater than* the F value in the table (or equal to it), you have a statistically significant reason to reject your null hypothesis.

Example:

In the table below we have four sets of data (five observations in each of four test situations)—don't worry about what the experiment was. The question here, as always, is

$$H_0: \overline{X}A = \overline{X}B = \overline{X}C = \overline{X}D.$$

We will work out the steps of the ANOVA test to see if there is enough difference between the means of these samples to permit us to reject the null hypothesis.

Raw data:

Sample A	Sample B	Sample C	Sample D
114 grams	119 grams	112 grams	117 grams
115 "	120 "	116 "	117 "
111 "	119 "	116 "	114 "
110 "	116 "	115 "	112 "
112 "	116 "	112 "	117 "

Step 1. $\Sigma x = 114 + 115 + \ldots 112 + 117 = 2{,}300.$

Step 2. $\Sigma x^2 = (114)^2 + (115)^2 + \ldots (112)^2 + (117)^2$
$= 264{,}652$

Step 3.

Sample A	Sample B	Sample C	Sample D
$\Sigma x_A = 562$	$\Sigma x_B = 590$	$\Sigma x_C = 571$	$\Sigma x_D = 577$
$\overline{X}_A = 112.4$	$\overline{X}_B = 118.0$	$\overline{X}_C = 114.2$	$\overline{X}_D = 115.4$
$n_A = 5$	$n_B = 5$	$n_C = 5$	$n_D = 5$

Step 4. $SS_T = \Sigma x^2 - \dfrac{(\Sigma x)^2}{N}$

N = 20 (5 observations in each of the 4 treatments)

$SS_T = 264{,}652 - \dfrac{(2{,}300)^2}{20} = 264{,}652 - 264{,}500 = 152$

Step 5.

$$SS_b = \left[\dfrac{(\Sigma x_A)^2}{n_A} + \dfrac{(\Sigma x_B)^2}{n_B} + \dfrac{(\Sigma x_C)^2}{n_C} + \dfrac{(\Sigma x_D)^2}{n_D} \right] - \dfrac{(\Sigma x)^2}{N}$$

$$SS_b = \left[\dfrac{(562)^2}{5} + \dfrac{(590)^2}{5} + \dfrac{(571)^2}{5} + \dfrac{(577)^2}{5} \right] - 264{,}500$$

$SS_b = (63{,}168.8 + 69{,}620 + 65{,}208.2 + 66{,}585.8) - 264{,}500$

$SS_b = 264{,}582.8 - 264{,}500 = 82.8$

Step 6. $SS_w = SS_T - SS_b \quad SS_w = 152 - 82.8 = 69.2$

Step 7.

$V_b = \dfrac{SS_b}{k-1}$ (there are 4 treatments so k – 1 = 3)

$V_b = \dfrac{82.8}{3} = 27.60$

Step 8.

$V_w = \dfrac{SS_w}{N-k}$ (N = 20, k = 4, so N – k = 16)

$V_w = \dfrac{69.2}{16} = 4.325$

Now that we have V_b and V_w,

$$F = \dfrac{27.6}{4.325} \quad \text{or} \quad F = 6.38.$$

Now that we have calculated the test statistic F, we need to consult the table of F values for the two different degrees of freedom (3 from the numerator and 16 from the denominator). Reading across the top of the table to 3, and down the side to 16, we find that the F value at the .05 alpha level of confidence is 3.24. Because our calculated F is much greater, we have ample reason to reject our null hypothesis.

Try this example:

Suppose an experimenter was interested in the importance of three sites in the brain controlling aggressive behavior. In the experiment, he tested four groups of rats. In three of the groups, he lesioned one of three different sites in the brain. The fourth group was a control receiving no lesions. Then, working within groups receiving similar treatment, he caged the rats in pairs and observed the number of aggressive encounters that occurred in two hours. Let us suppose these are his data. (Data are taken from Terrace and Parker; see bibliography.)

Number of fights in two hours			
Group A lesion site 1	Group B lesion site 2	Group C lesion site 3	Group D control-no lesion
12	12	19	9
14	16	18	5
8	15	25	6
11	13	21	8
15	19	22	3
12			5
10			
14			

Table of critical values for the statistic F calculated in the ANOVA test.[1]
The alpha level is .05.

Degrees of freedom for the greater variance (usually the numerator)

	1	2	3	4	5	6	7	8	9	10	12	15	20	24	30	40	60	120	∞
1	161.4	199.5	215.7	224.6	230.2	234.0	236.8	238.9	240.5	241.9	243.9	245.9	248.0	249.1	250.1	251.1	252.2	253.3	254.3
2	18.51	19.00	19.16	19.25	19.30	19.33	19.35	19.37	19.38	19.40	19.41	19.43	19.45	19.45	19.46	19.47	19.48	19.49	19.50
3	10.13	9.55	9.28	9.12	9.01	8.94	8.89	8.85	8.81	8.79	8.74	8.70	8.66	8.64	8.62	8.59	8.57	8.55	8.53
4	7.71	6.94	6.59	6.39	6.26	6.16	6.09	6.04	6.00	5.96	5.91	5.86	5.80	5.77	5.75	5.72	5.69	5.66	5.63
5	6.61	5.79	5.41	5.19	5.05	4.95	4.88	4.82	4.77	4.74	4.68	4.62	4.56	4.53	4.50	4.46	4.43	4.40	4.36
6	5.99	5.14	4.76	4.53	4.39	4.28	4.21	4.15	4.10	4.06	4.00	3.94	3.87	3.84	3.81	3.77	3.74	3.70	3.67
7	5.59	4.74	4.35	4.12	3.97	3.87	3.79	3.73	3.68	3.64	3.57	3.51	3.44	3.41	3.38	3.34	3.30	3.27	3.23
8	5.32	4.46	4.07	3.84	3.69	3.58	3.50	3.44	3.39	3.35	3.28	3.22	3.15	3.12	3.08	3.04	3.01	2.97	2.93
9	5.12	4.26	3.86	3.63	3.48	3.37	3.29	3.23	3.18	3.14	3.07	3.01	2.94	2.90	2.86	2.83	2.79	2.75	2.71
10	4.96	4.10	3.71	3.48	3.33	3.22	3.14	3.07	3.02	2.98	2.91	2.85	2.77	2.74	2.70	2.66	2.62	2.58	2.54
11	4.84	3.98	3.59	3.36	3.20	3.09	3.01	2.95	2.90	2.85	2.79	2.72	2.65	2.61	2.57	2.53	2.49	2.45	2.40
12	4.75	3.89	3.49	3.26	3.11	3.00	2.91	2.85	2.80	2.75	2.69	2.62	2.54	2.51	2.47	2.43	2.38	2.34	2.30
13	4.67	3.81	3.41	3.18	3.03	2.92	2.83	2.77	2.71	2.67	2.60	2.53	2.46	2.42	2.38	2.34	2.30	2.25	2.21
14	4.60	3.74	3.34	3.11	2.96	2.85	2.76	2.70	2.65	2.60	2.53	2.46	2.39	2.35	2.31	2.27	2.22	2.18	2.13
15	4.54	3.68	3.29	3.06	2.90	2.79	2.71	2.64	2.59	2.54	2.48	2.40	2.33	2.29	2.25	2.20	2.16	2.11	2.07
16	4.49	3.63	3.24	3.01	2.85	2.74	2.66	2.59	2.54	2.49	2.42	2.35	2.28	2.24	2.19	2.15	2.11	2.06	2.01
17	4.45	3.59	3.20	2.96	2.81	2.70	2.61	2.55	2.49	2.45	2.38	2.31	2.23	2.19	2.15	2.10	2.06	2.01	1.96
18	4.41	3.55	3.16	2.93	2.77	2.66	2.58	2.51	2.46	2.41	2.34	2.27	2.19	2.15	2.11	2.06	2.02	1.97	1.92
19	4.38	3.52	3.13	2.90	2.74	2.63	2.54	2.48	2.42	2.38	2.31	2.23	2.16	2.11	2.07	2.03	1.98	1.93	1.88
20	4.35	3.49	3.10	2.87	2.71	2.60	2.51	2.45	2.39	2.35	2.28	2.20	2.12	2.08	2.04	1.99	1.95	1.90	1.84
21	4.32	3.47	3.07	2.84	2.68	2.57	2.49	2.42	2.37	2.32	2.25	2.18	2.10	2.05	2.01	1.96	1.92	1.87	1.81
22	4.30	3.44	3.05	2.82	2.66	2.55	2.46	2.40	2.34	2.30	2.23	2.15	2.07	2.03	1.98	1.94	1.89	1.84	1.78
23	4.28	3.42	3.03	2.80	2.64	2.53	2.44	2.37	2.32	2.27	2.20	2.13	2.05	2.01	1.96	1.91	1.86	1.81	1.76
24	4.26	3.40	3.01	2.78	2.62	2.51	2.42	2.36	2.30	2.25	2.18	2.11	2.03	1.98	1.94	1.89	1.84	1.79	1.73
25	4.24	3.39	2.99	2.76	2.60	2.49	2.40	2.34	2.28	2.24	2.16	2.09	2.01	1.96	1.92	1.87	1.82	1.77	1.71
26	4.23	3.37	2.98	2.74	2.59	2.47	2.39	2.32	2.27	2.22	2.15	2.07	1.99	1.95	1.90	1.85	1.80	1.75	1.69
27	4.21	3.35	2.96	2.73	2.57	2.46	2.37	2.31	2.25	2.20	2.13	2.06	1.97	1.93	1.88	1.84	1.79	1.73	1.67
28	4.20	3.34	2.95	2.71	2.56	2.45	2.36	2.29	2.24	2.19	2.12	2.04	1.96	1.91	1.87	1.82	1.77	1.71	1.65
29	4.18	3.33	2.93	2.70	2.55	2.43	2.35	2.28	2.22	2.18	2.10	2.03	1.94	1.90	1.85	1.81	1.75	1.70	1.64
30	4.17	3.32	2.92	2.69	2.53	2.42	2.33	2.27	2.21	2.16	2.09	2.01	1.93	1.89	1.84	1.79	1.74	1.68	1.62
40	4.08	3.23	2.84	2.61	2.45	2.34	2.25	2.18	2.12	2.08	2.00	1.92	1.84	1.79	1.74	1.69	1.64	1.58	1.51
60	4.00	3.15	2.76	2.53	2.37	2.25	2.17	2.10	2.04	1.99	1.92	1.84	1.75	1.70	1.65	1.59	1.53	1.47	1.39
120	3.92	3.07	2.68	2.45	2.29	2.17	2.09	2.02	1.96	1.91	1.83	1.75	1.66	1.61	1.55	1.50	1.43	1.35	1.25
∞	3.84	3.00	2.60	2.37	2.21	2.10	2.01	1.94	1.88	1.83	1.75	1.67	1.57	1.52	1.46	1.39	1.32	1.22	1.00

Degrees of freedom for the lesser variance (usually the denominator)

1. Adapted from R.G.D. Steel and J.H. Torrie. *Principles and Procedures of Statistics*. McGraw-Hill Book Company, New York. 1960.

The Friedman Test

Purpose:

The Friedman Test, a non-parametric test of differences between means, tests for significant differences between the responses of several matched samples (paired samples) exposed to three or more treatments. This experimental design is often called a randomized block design. The Friedman Test is the non-parametric equivalent of the ANOVA test.

Warnings and precautions:

1. Data must be measured on an ordinal, interval, or ratio scale.
2. The items in each sample must be of a matched design. Hence the same number of items will be in each group and must be randomly assigned to each treatment (see the example).

Procedure:

The null hypothesis to be tested is that the means of the responses to three or more treatments do not differ significantly from each other.

(H_0: $\overline{X}A = \overline{X}B = \overline{X}C$...etc.)

1. Place the response values (or scores) in a two-way table that has k columns (k = the number of samples or treatments) and b rows or blocks (b = the number of observations or samples in each treatment).
2. Rank the values in each row from 1 to however many columns you have. If this is not clear, see the example and also the discussion of ranking data in the Spearman's Rank Correlation Test. Tied values each receive the average rank of the ranks that would have been assigned to those tied values.
3. Add up the ranks found in each column producing R_A (the sum of the ranks in column or treatment A), R_B, R_C, ...etc.
4. Calculate the degrees of freedom according to the formula, df = k - 1 (k = the number of treatments or columns).

5. Calculate the value of the test statistic according to this formula:

$$X_r^2 = \frac{12}{bk(k + 1)} \quad \Sigma\,(R_s)^2 \; - \; 3b(k + 1)$$

b = the number of rows or observations
k = the number of columns or treatments

$\Sigma(R_s)^2$ = Take the sum of each column's ranks and square it. Add all the squared totals of the ranks of each column.

6. For small values of b and k, use the accompanying table to check your calculated test statistic against the critical values. For values not found on this table, use the χ^2 table found on page 95.
7. If the calculated value of X_r^2 is equal to or greater than the critical value in the table, you may reject your null hypothesis and accept the alternative hypothesis—that the means of the responses to the treatments are *not* equal. If the calculated test statistic is less than the critical value on the table you are using, you fail to reject your null hypothesis.

Example:

An experimenter wishes to study potential stress in mice (as measured by weight loss) caused by smoke inhalation, exposure to bright light and exposure to rock music. He takes nine cages, each containing three mice which are littermates of the same age and sex, and randomly assigns one of the three mice from each cage to each of the following treatment groups:

Treatment A—mice placed in a chamber full of cigar smoke for one hour each day.

Treatment B—mice placed in very bright light for one hour each day.

Treatment C—mice exposed to rock music for one hour each
day.

Except for the time when the mice are exposed to the test
conditions, the mice are maintained in the same fashion.

The following table shows the weight loss in grams for each of
the mice exposed to each treatment.

Cage #	Treatment A (smoke)		Treatment B (light)		Treatment C (music)	
	wt. loss	rank	wt. loss	rank	wt. loss	rank
1	5	(2)	4	(1)	6	(3)
2	2.5	(3)	1	(1)	2	(2)
3	4	(1)	5	(2)	7	(3)
4	2	(2)	1	(1)	3	(3)
5	4	(2)	3	(1)	5.5	(3)
6	1	(1.5)	1	(1.5)	2	(3)
7	4	(2)	3.5	(1)	5	(3)
8	4	(1)	5	(3)	4.5	(2)
9	1.5	(1)	2	(2)	3	(3)
sum of the ranks		15.5		13.5		25

$H_0 : \bar{X}A = \bar{X}B = \bar{X}C$

calculations:

$$X_r^2 = \frac{12}{bk(k+1)} \; \Sigma \, (R_s)^2 - 3b(k+1)$$

$$X_r^2 = \frac{12}{(9)\,(3)\,(3+1)}[\,(15.5)^2+(13.5)^2+(25)^2\,]-(3)\,(9)\,(3+1)$$

$$X_r^2 = 8.39$$

Using the table on page 62, the critical value for this example,
with a **b** of 9 and a **k** of 3, is 6.222. Because our calculated test
statistic X_r^2 is 8.39, which is greater than the critical value of
6.222, we are able to reject our null hypothesis in favor of the

alternative hypothesis. (Note that you have only shown that the means are unequal—not which treatment causes this to be so.)

Try this example:

In a test comparing three commercial fertilizers, four kernels of corn were randomly selected from each of 12 ears of Illini Chief corn. One of each set of four was subjected to each of the following experimental situations. Group 1 was planted in soil with no added fertilizer, group 2 was planted in similar soil with Fertilizer A, group 3 with Fertilizer B, and group 4 with Fertilizer C. All kernels were planted on the same day and subjected to identical environmental conditions. After two weeks, the growth in centimeters was measured for each young plant. The data looked like this:

Growth in Centimeters

Ear number	No fertilizer	Fertilizer "A"	Fertilizer "B"	Fertilizer "C"
1	10cm	12cm	13 cm	11cm
2	11.5	11	14	10
3	9	16	17	15
4	12	12	14	16
5	15	18	17	16
6	16	20	22	24
7	11	21	23	24
8	10.5	14	18	16
9	9	20	22	10
10	14	15	11	17
11	13	25	20	18
12	12	14	15	16

Is there a statistical reason to reject the null hypothesis that the mean growth under the four experimental situations is the same? (Note that with this larger sample size you will have to use the χ^2 table found on page 95.)

Table of critical values for the Friedman Test at the .05 alpha level.[1]

k	b	Critical Value
3	3	6.000
3	4	6.500
3	5	6.400
3	6	7.000
3	7	7.143
3	8	6.250
3	9	6.222
4	2	6.000
4	3	7.400
4	4	7.800

(for larger sample sizes, use the χ^2 table on page 95, with k-1 d.f.)

1. M. Friedman, 1937. The use of ranks to avoid the assumption of normality implicit in the analysis of variance. *J. Am. Stat. Assoc.* 32:675–701.

The Kruskal-Wallis Test

Purpose:

The Kruskal-Wallis Test is a non-parametric test for differences between means that can be used in any situation appropriate for single factor analysis of variance. However, it does not have to meet all the requirements of a parametric test. It is appropriate for comparing more than two independent samples of equal or unequal size.

Warnings and precautions:

1. The measurement scale must be an ordinal, interval, or ratio scale.
2. All random variables should be continuous.
3. A large number of tied values may distort the true level of significance.
4. The more-than-two samples must be independent of each other and the values within each sample must also be independent.

Notation:

k = the number of samples (the number of groups compared)

n_1 = the number of items in the first sample

n_2 = the number of items in the second sample

n_3 = the number of items in the third sample, etc.

N = the total number of observations in all samples ($n_1 + n_2 + n_3 + n_4 + n_5 +$ etc.)

R_1 = the sum of the ranks assigned to the values in the first sample

R_2 = the sum of the ranks assigned to the values in the second sample, etc.

H = the test statistic you will calculate

The null hypothesis:

The means of the k samples do not differ significantly from each other. ($\overline{X}_1 = \overline{X}_2 = \overline{X}_3 = \overline{X}_4$ etc.)

The alternative hypothesis is that at least one of the samples contains values that are significantly different from the others. (Note that with this test you conclude *only* that the samples are unequal—you do not determine which one or more is different.)

Procedure:

1. Rank all of the N observations into a single series by assigning the rank of 1 to the smallest value, etc. Rank without regard to group. Ties are assigned the average value of the ranks that would have been assigned to them. (See the example and also the discussion of ranking data in the Spearman's Rank Correlation test.)
2. Calculate R for each of the k samples.
3. Calculate the Kruskal-Wallis test statistic H by the following formula:

$$H = \frac{12}{N(N+1)} \left[\frac{R_1^2}{n_1} + \frac{R_2^2}{n_2} + \frac{R_3^2}{n_3} \text{ etc.} \right] - 3(N+1)$$

4. If k = 3, *and* all sample sizes are 5 or less, use the Kruskal-Wallis test table for comparing your calculated H to the critical value.

If k = 4 or more, *or* the sample sizes are greater than 5, use the χ^2 table with k-1 degrees of freedom (page 95).

If the test statistic H is equal to or greater than the critical value in either table, you may reject your null hypothesis.

Example:

Four farmers each claim to have developed a better way of raising corn. They decide to use the number of ounces of dried corn kernels per plant as a measure of yield. Each farmer picks one good plant from each row for his measurement. The farms differ in number of rows. The table of data on the next page presents the ounces per plant and the rank for each value.

FARM A weights	rank	FARM B weights	rank	FARM C weights	rank	FARM D weights	rank
28	4	63	35	63	35	63	35
32	7	55	28.5	48	23	69	40
25	1	66	38	26	2	56	30
36	12	58	31	33	8.5	61	33
36	12	67	39	41	17.5	45	21
36	12	55	28.5	37	14	52	26
40	16	47	22	29	5	59	32
33	8.5	51	25	34	10	53	27
50	24	42	19	30	6	65	37
44	20			38	15		
				27	3		
				41	17.5		

$R_1 = 116.5$ $R_2 = 266$ $R_3 = 156.5$ $R_4 = 281$

$n_1 = 10$ $n_2 = 9$ $n_3 = 12$ $n_4 = 9$

$N = 40$

$k = 4$

degrees of freedom $= 3$

This is the calculation of the test statistic H:

$$H = \frac{12}{40(40 + 1)}\left[\frac{116.5^2}{10} + \frac{266^2}{9} + \frac{156.5^2}{12} + \frac{281^2}{9}\right] - 3(40 + 1)$$

or:

$$H = \frac{12}{1640}\left[\frac{13572.25}{10} + \frac{70756}{9} + \frac{24492.25}{12} + \frac{78961}{9}\right] - 123$$

or:

$H = .007317(1357.225 + 7861.78 + 2041.02 + 8773.44) - 123$

$H = .007317 (20033.465) - 123$ and finally,

$H = 23.59$

When the H of 23.59 is compared to the critical value on the χ^2 table at 3 degrees of freedom, we find 23.59 is larger than 7.81 and therefore we may reject our null hypothesis. In this case, we may conclude only that one or more of the farmer's samples of corn are significantly different from the others. We can gain some idea of how the samples differ by calculating an average weight (note that the sample sizes are unequal) but this statistical test does not allow us to draw any conclusions about how they differ. Other statistical tests exist which permit you to discriminate more finely.

(In our example we had more than three samples and more than five values in each sample, so we compared our test statistic H to critical values in the χ^2 table. If there had been only three samples and the sample sizes had been five or less, we would have used the table of critical values given on page 67.)

Try this example:

Three first grade teachers, who use different teaching techniques, decided to compare their methods for teaching reading. Each selected a test group of students who could not read at the beginning of the school year, taught them in their usual way, and then gave them a standard reading test at the end of the year. (They assumed all the test students had an equal learning ability.)

Teacher A's students had the following test scores:
60 75 72 63 50 80 88 90 ($n_1 = 8$, average score $= 72.25$)
Teacher B's students had the following scores:
55 65 82 86 90 92 60 58 95 70 ($n_2 = 10$, average score $= 75.3$)
Teacher C's students had these scores:
48 52 92 85 60 62 65 75 80 58 76 78 ($n_3 = 12$, average score $= 69.25$)

Is there any statistical reason, based on these test scores, to believe one teaching method is any better than another? The null hypothesis is that there is no measurable difference in performance among the students taught in the different methods.

$$H_0 : \overline{X}A = \overline{X}B = \overline{X}C$$

Table of critical values at the .05 level for the Kruskal-Wallis Test[1]

Sample Sizes

n_1	n_2	n_3	Critical Values
3	2	2	4.714
3	3	1	5.143
3	3	2	5.361
3	3	3	5.600
4	2	1	4.8214*
4	2	2	5.333
4	3	1	5.208
4	3	2	5.444
4	3	3	5.727
4	4	1	4.967
4	4	2	5.455
4	4	3	5.598
4	4	4	5.692
5	2	1	5.000
5	2	2	5.160
5	3	1	4.960
5	3	2	5.251
5	3	3	5.648
5	4	1	4.986
5	4	2	5.273
5	4	3	5.656
5	4	4	5.657
5	5	1	5.127
5	5	2	5.338
5	5	3	5.705
5	5	4	5.666
5	5	5	5.780

(*this is actually at .057, no alpha level of .05 was available at this sample size)

1. Adapted from J.H. Zar, *Biostatistical Analysis*. Prentice-Hall, Inc. Englewood Cliffs, N.J. 1974.

The Rank Sum Test

(based on the White modification of the Wilcoxon Rank Sum Test)

Purpose:

The Rank Sum Test, a non-parametric test of differences between means, is used to test for significant differences between two samples of equal or unequal size.

Warnings and precautions:

1. This test is not appropriate for paired data.
2. Data must be measured on an ordinal, interval or ratio scale.

Procedure:

Null hypothesis: The mean of sample A is not significantly different from that of sample B ($H_0 : \overline{X}A = \overline{X}B$).

1. Rank all data points from both samples into a single series. Tied absolute values each get the average of the ranks that they would have been assigned if the values had not been tied.
2. If the sample sizes are equal, obtain the sum of the ranks for each sample. Call the smaller sum "T" and go to step 4. If the sample sizes are not equal, go to step 3.
3. Call the smaller sample size "n_1" and the larger "n_2." Obtain the rank sum of the smaller sample (n_1) by adding the ranks of its points. Call this sum T_1. Calculate T_2 by this formula:

$T_2 = n_1 (n_1 + n_2 + 1) - T_1$ where n_1 = the sample size of the smaller sample, and n_2 = the sample size of the larger sample. Let T = the smaller of T_1 or T_2.

4. For sample sizes not included on the table on page 71, go to step 5. For sample sizes included in the table, compare T with the critical value in the table. If the smaller rank sum is equal to, or *smaller* than, the critical value, you may reject the null hypothesis. If the value is greater than the critical value, you fail to reject the H_0.

68

5. Compute the statistic Z by the following formula:

$$Z = \frac{(|m - T| - \frac{1}{2})}{S}$$

$$m = \frac{n_1(n_1 + n_2 + 1)}{2} \quad (n_1 \text{ and } n_2 \text{ are as above})$$

$$S = \sqrt{\frac{n_2 m}{6}}$$

T = the smaller of the two rank sums (T_1 or T_2)

If Z is *equal to* or *greater than* 1.96, you may reject your null hypothesis. If not, you fail to reject the H_0.

Example:

A paperboy covers a morning route to the south of town and an evening route to the north. The sizes of the dogs that attack the paperboy in the morning route are listed in Sample A and the sizes of those attacking in the evening are in Sample B. Is the paperboy attacked by larger dogs on one route than on the other? The null hypothesis would be that the average size of the dogs attacking on the southern route was equal to the average size of the dogs attacking on the northern route. (H_0: $\overline{X}A = \overline{X}B$). In the following table we have shown how you would list your size measurements (in the order you took them) for each sample and then assign ranks to the measurements of both groups treated as one. (Rank 1 is assigned to the smallest measurement, etc.)

Sample A	Rank	Sample B	Rank
12"	3	33"	15
14"	4	21"	7
18"	5.5	23"	8.5
25"	10	23"	8.5
28"	12	46"	19
10"	2	35"	16
7"	1	31"	14
27"	11	37"	17
30"	13	42"	18
		18"	5.5
sum of ranks	61.5	sum of ranks	128.5

Sample A has only nine readings; it is the smaller sample size, or n_1. T_1 is the sum of the ranks of the smaller sample (61.5). T_2 = the first sample size, n_1 (or 9), times the sum of the two sample sizes (19) plus 1, or 9 x 20, less T_1. 9 x 20 = 180. 180 - 61.5 = 118.5. When T_1 (which is 61.5) is compared with T_2 (118.5), we find that T_1 is the *smaller* and therefore is the figure T which we must compare with the critical value for our two sample sizes on the chart on page 71. Reading across the top line to our n_1 or 9, and down the side to our n_2 which is 10, we find the critical value is 65. Our T (61.5) is smaller than the critical value 65; therefore we may reject our null hypothesis that the samples were equal.

Try this example:

To examine the potential ability of vitamin C to prevent colds, an experimental group of 15 teenagers (selected at random from a group of 30) were given 10 g of vitamin C per day for one year. A control group of the remaining 15 were given similar-looking, harmless pills which contained no vitamin supplement. In other respects, each youth maintained his normal diet. Each recorded the number of colds he had during the test year.

These are the results from the experimental group receiving vitamin C (n_1 = 15):
  4 2 3 1 5 4 0 2 1 6 0 2 3 1 2 (an average of 2.4 colds/yr.)

These are the numbers of colds recorded by the control group (n_2 = 15):
  4 5 2 6 8 7 10 9 5 4 6 1 1 3 4 (average number of colds = 5)

Is there a statistically significant reason to conclude that vitamin C is effective in cold prevention? $H_0 : \overline{X}_{exp.} = \overline{X}_{contr.}$

Table of critical values for the Rank Sum Test at the .05 alpha level.[1]

sample sizes for n_1

sample sizes for n_2

n_2	2	3	4	5	6	7	8	9	10	11	12	13	14	15
4			10											
5		6	11	17										
6		7	12	18	26									
7		7	13	20	27	36								
8	3	8	14	21	29	38	49							
9	3	8	15	22	31	40	51	63						
10	3	9	15	23	32	42	53	65	78					
11	4	9	16	24	34	44	55	68	81	96				
12	4	10	17	26	35	46	58	71	85	99	115			
13	4	10	18	27	37	48	60	73	88	103	119	137		
14	4	11	19	28	38	50	63	76	91	106	123	141	160	
15	4	11	20	29	40	52	65	79	94	110	127	145	164	185
16	4	12	21	31	42	54	67	82	97	114	131	150	169	
17	5	12	21	32	43	56	70	84	100	117	135	154		
18	5	13	22	33	45	58	72	87	103	121	139			
19	5	13	23	34	46	60	74	90	107	124				
20	5	14	24	35	48	62	77	93	110					
21	6	14	25	37	50	64	79	95						
22	6	15	26	38	51	66	82							
23	6	15	27	39	53	68								
24	6	16	28	40	55									
25	6	16	28	42										
26	7	17	29											
27	7	17												
28	7													

1. Adapted from R.G.D. Steel and J.H. Torrie. *Principles and Procedures of Statistics*. McGraw-Hill Book Co., N.Y. 1960.

Regression Analysis
(Simple Linear Regression)[1]

Purpose:

The Regression Analysis tests to see whether there is a functional relationship between variables. In simple linear regression, we are dealing with only two variables and testing to see whether the functional relationship between the dependent variable Y and the independent variable X can be described as a straight line. This type of analysis can be used to examine causal relationships between variables and to predict one variable given the value of the other.

The hypothesis to be tested is whether the slope of the line in the equation $Y = a + bX$ is, for statistical purposes, equal to 0. $H_0 : b = 0$.

The regression equation $Y = a + bX$ describes a relationship in which Y is the dependent variable, X is the independent variable, and b is the slope of the line which describes the change in Y per unit change in X. And "a" is the point where the line crosses the Y axis, or the "Y intercept." This relationship is illustrated in this figure:

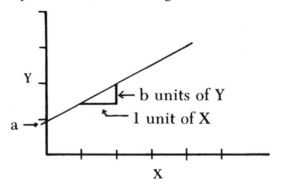

1. The authors thank Nicholas J. Cheper, Department of Zoology, University of Tennessee/Knoxville, for contributing this test.

Warnings and Precautions:

1. The test assumes the values of X, the independent variable, are measured without error. (The values of X are fixed by the experimenter and only the values of Y are free to vary.)
2. The Y values must be measured on a continuous scale. The X values can be discrete or continuous but usually are discrete.
3. The test assumes that values of Y are from a normal distribution.
4. It also assumes the variance around the regression line is the same for all XY pairs.

Notation:

ΣX = the sum of the values for X, or the independent variable
ΣX^2 = the sum of the squares of each independent variable
ΣY = the sum of the values for Y, the dependent variable
ΣY^2 = the sum of the squares of the dependent variables
ΣXY = multiply each X by each Y and add up the products of each XY pair
n = the number of XY pairs
SS_T = the total sum of the squares of the Y values
SS_r = the sum of the squares due to regression
SS_E = the residual sum of the squares
V_r = the variance due to regression
V_E = the variance due to residuals

Procedure:

Step 1. Arrange the data by XY pairs (see the example if this isn't clear).

Step 2. Compute ΣX by adding up all the X values, and compute the $\overline{X}$ (the mean of X) by dividing the value just calculated by the number of X readings.

Step 3. Compute ΣX^2 by squaring each X reading and adding the squares.

Step 4. Compute ΣY and $\overline{Y}$ (the mean of Y) as you did for the X values.

Step 5. Compute ΣY^2.

Step 6. Compute ΣXY.

Step 7. Compute the slope b by the following formula:

$$b = \frac{\Sigma XY - \dfrac{(\Sigma X)(\Sigma Y)}{n}}{\Sigma X^2 - \dfrac{(\Sigma X)^2}{n}}$$

Step 8. Compute a, the Y intercept where $X = 0$, by the following formula: $\qquad a = \overline{Y} - b\overline{X}$

Step 9. Place the values you have calculated into this equation: $\qquad Y = a + bX$

Step 10. Compute SS_T by the formula:

$$SS_T = \Sigma Y^2 - \frac{(\Sigma Y)^2}{n}$$

Step 11. Compute SS_r by the formula:

$$SS_r = b\left[\Sigma XY - \frac{(\Sigma X)(\Sigma Y)}{n}\right]$$

Step 12. Subtract SS_r from SS_T to get the SS_E.

Step 13. Now we need to divide the SS_r and the SS_E by the appropriate degrees of freedom to get the variance for each (the V_r and the V_E). The degrees of freedom for SS_r is *always* 1. Therefore, $SS_r = V_r$. The degrees of freedom for $SS_E = n - 2$. Therefore the formula for the variance due to residuals is

$$V_E = \frac{SS_E}{n - 2}$$

Step 14. The last step is to set up an F ratio using the formula

$$F = \frac{V_r}{V_E}$$

and calculate a value for F which can be compared with the critical values for F in the table used with the ANOVA test on page 57. The critical value for F must be found for 1 degree of freedom in the numerator and n - 2 degrees of freedom in the denominator. If the F value you have calculated is *equal to* or *greater than* the critical value in the F table, you may reject the null hypothesis that b = 0 and accept the alternate hypothesis. Your data can be explained by a simple linear regression with a slope of b describing the functional linear relationship between your values of X and those of Y. If the $F_{calculated}$ is less than the $F_{critical}$ on the table, you fail to reject your null hypothesis and must conclude that no linear relationship exists between your two variables.

Example:

Let us suppose that you wanted to know whether the length and girth of bean shoots were related. Here are some measurements of the circumference of bean shoots (the Y values) measured for various lengths of bean shoots (the X values).

X — shoot length	Y — shoot girth
1	2.5
3	4.6
4	9.5
7	11.6
9	14.8
11	20.0

Step 1. The data are already arranged in XY pairs.

Step 2. $\Sigma X = 35$ $\overline{X} = 5.83$

Step 3. $\Sigma X^2 = 277$

Step 4. $\Sigma Y = 63$ $\overline{Y} = 10.5$

Step 5. $\Sigma Y^2 = 871.26$

Step 6. $\Sigma XY = 488.7$

Step 7. Inserting the figures just calculated into the formula for calculating the slope, b, we get the following equation:

$$b = \cfrac{488.7 - \cfrac{(35)\ (63)}{6}}{277 - \cfrac{(35)^2}{6}} \quad \text{or } b = \frac{488.7 - 367.5}{277 - 204.0}$$

$$b = \frac{121.2}{73} \quad \text{or } b = 1.664$$

Step 8. Calculate a, the Y intercept, by the formula

$a = \overline{Y} - bX.$

$a = 10.5 - (1.664)\ (5.83)$ or $a = .80$

Step 9. $Y = a + bX$, so $Y = .80 + 1.664X$

Step 10. Calculate SS_T. $SS_T = 817.26 - \dfrac{(63)^2}{6}$

$SS_T = 209.76$

Step 11. Calculate SS_r.

$$SS_r = 1.664 \left[488.7 - \frac{(35)\ (63)}{6} \right] \quad \text{or } SS_r = (1.664)\ (121.2)$$

$SS_r = 201.6768$

Step 12. Calculate SS_E. $SS_E = 209.76 - 201.6768$

$SS_E = 8.0832$

Step 13. Calculate V_r and V_E.

$V_r = SS_r$ or 201.6768

$V_E = \dfrac{8.0832}{4}$ or $V_E = 2.0208$

Therefore the F ratio is

$$F = \frac{201.6768}{2.0208} \quad \text{or } F = 99.80$$

When we compare our calculated F of 99.80 to the appropriate F value on the F table (with the ANOVA test, page 57) at the appropriate degrees of freedom 1 and 4 ($n = 6$, $n - 2 = 4$), we find our $F_{calculated}$ of 99.8 is much greater than the $F_{critical}$ of 7.71. Therefore we can reject our null hypothesis that $b = 0$ (or that no linear functional relationship exists between the height and girth of bean shoots) and conclude that a dependent relationship exists.

Try this example:

The wing lengths of sparrows were measured at various times after hatching. The data are given below, with X being the age in days after hatching and Y the wing length measured in centimeters. Is there a linear functional relationship between the values of X and Y in the following set of data (taken from Zar, see bibliography)?

Age in days X	Wing length in cm. Y
3	1.4
5	2.2
6	2.4
10	3.2
14	4.7
17	5.0

The Sign Test

Purpose:

The Sign Test, a non-parametric test of differences between means (more correctly, medians), is used to determine whether two related samples differ significantly, or whether some treatment causes a change in an experimental group.

Warnings and precautions:

1. The data must be measured on an ordinal, interval or ratio scale.
2. The "experiment" may be of a related sample paired design.
3. The measured variable must have some sequential arrangement of magnitude.

Procedure:

Total the number of "greater than" responses. Total the number of "less than" responses. "No change" responses are ignored. Find the square on the Sign Test figure (page 80) that corresponds to these values. If this square lies in the "do not reject" region, you fail to reject your hypothesis. If it lies outside this region (in either reject region), you may reject your null hypothesis.

Example:

You measure the jumping abilities of 25 frogs. Then you give them all a dose of "Calaveras Tonic" and find that 20 of the frogs now can jump farther than they did before, and 5 jump less far. H_0: the average jumping distance before Calaveras Tonic = the average jumping distance after the tonic for each frog. You would find the Plus 20 (20 "more than" responses)/Minus 5 square. It lies in the upper "reject" region. You may reject your hypothesis and conclude that Calaveras Tonic makes frogs jump farther.

In the Sign Test, the H_0 is: the ratio of "more than" to "less than" responses is not significantly different from 50:50.

Try this example:

The strength of the knee-jerk reflex, measured in degrees of arc, was tested on 10 men of the same age under two conditions— with the muscles tensed (T) and relaxed (R). The data (taken from Guilford; see bibliography) are presented in the table below. The null hypothesis, when testing these data with the Sign Test, is they represent a random sample from the *same* population—or that there is no difference in the reflex response under the two test conditions. The signs (+ or -) in the table represent the changes between the tensed and relaxed conditions.

Knee-jerk reflexes, in degrees of arc, for 10 men tested under tensed (T) and relaxed (R) conditions.

T	R	sign of T − R
19	14	+
19	19	o
26	30	−
15	7	+
18	13	+
30	20	+
18	17	+
30	29	+
26	18	+
28	21	+

(Note that there are 10 pairs of observations but, in one pair, no change is involved. Therefore we have 8 pluses and 1 minus.)

Table for the Sign Test at the alpha level of .05.[1]

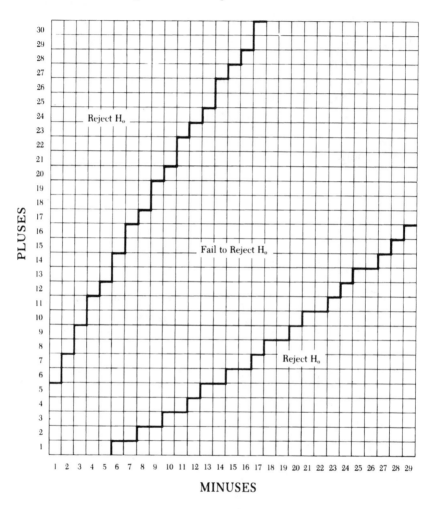

PLUSES

MINUSES

1. Constructed from values found in The Chemical Rubber Co., *Handbook of Probability and Statistics*, 2nd ed. (W.H. Beyer, ed.) The Chemical Rubber Company, Cleveland, Ohio. 1968.

Spearman's Rank Correlation

Purpose:

This non-parametric test of relationships may be used to establish whether or not two variables are correlated.

Warnings and precautions:

1. The data must be from an ordinal, interval or ratio scale of measurement.
2. The individual data points must be independent.
3. Remember that correlation does not necessarily imply cause and effect.

Procedure:

The test statistic, is indicated by "r_s." H_0: there is no correlation between the two variables in question.

1. Separately rank the variables for each data point within the two groups. Tied absolute values each get the average rank of these two values had they not been tied. If this is unclear, look at the absolute values of some data and how they are ranked:

values:	8	12	14	16	16	20	30
ranks:	1	2	3	4.5	4.5	6	7

(here two values are tied)

or,

values:	8	12	16	16	16	20	30
ranks:	1	2	4	4	4	6	7

(three are tied)

2. Compute the differences between the ranks (d) for the two variables for each data point.
3. Square each difference.
4. Sum the square of the differences (Σd^2).
5. Apply the following formula:

$$r_s = 1 - \left(\frac{6\,\Sigma\,d^2}{n^3 - n} \right)$$

where d^2 = the square of the differences between the ranks for the two variables that establish each point, and n = the number of individual points.

    6. Compare the calculated statistic r_s with the critical value given in the table on page 84 for the appropriate sample size.

If r_s is *greater than*, or *equal to*, the critical value, you may reject the null hypothesis.

Example:

Suppose that you want to know whether a correlation exists between heights and weights of the six men listed in the following table. Your H_0 would be that no correlation exists between the height and the weight of these men. This is how you would arrange your data and rank the variables:

Individual	Height (inches)	Rank	Weight (lbs)	Rank	d	d²
Fred	68	2	140	2	0	0
Ralph	69	3	170	5	2	4
Sam	71	4	160	4	0	0
Tom	73	5	150	3	2	4
Dick	66	1	130	1	0	0
Harry	74	6	180	6	0	0
					Σd^2 =	8

This would be your calculation:

$$r_s = 1 - \left(\frac{6\Sigma d^2}{n^3 - n} \right) \qquad r_s = 1 - \left(\frac{48}{6^3 - 6} \right)$$

$$r_s = 1 - \left(\frac{48}{210} \right) \quad r_s = 1 - .23 \quad \text{therefore} \quad r_s = .77$$

You have now calculated r_s, which is .77. Go to the rank correlation table (which we have given for the alpha level of .05 only) and find the critical value for your sample size (n) which is 6. The critical value from the table is .886. Since your calculated r_s is less than the critical value, the correlation is not significant at the .05 alpha level of confidence. You have failed to reject your null hypothesis. You cannot assume a relationship exists between the height and the weight of these men.

Try this example:

Fifteen individuals were shown sets of limericks and sets of cartoons and asked to judge the humor value of each on a 5 point scale. In the data given in the following table, the cartoon score is the sum of the points assigned to each item in the set by one individual. The limerick score also represents that same individual's total scores for the limericks in the test set. Because there are several tied scores, we have supplied the ranks of each set for you. (The data are taken from Guilford; see bibliography.)

Cartoon score	Limerick score	R_1	R_2	d	d^2
47	75	11	8	3	9
71	79	4	6	2	4
52	85	9	5		
48	50	10	14	etc.	
35	49	14.5	15		
35	59	14.5	12		
41	75	12.5	8		
82	91	1	3		
72	102	3	1		
56	87	7	4		
59	70	6	10		
73	92	2	2		
60	54	5	13		
55	75	8	8		
41	68	12.5	11		

In this instance the Spearman's Rank Correlation Test would determine whether a statistically significant correlation existed between the perceived humor content in the cartoons and the limericks. The null hypothesis would be that no correlation existed between the two sets of data.

Table of critical values for different sample sizes at the .05 alpha level to be used with the Spearman's Rank Correlation test.[1]   (n = sample size)

n	critical value	n	critical value	n	critical value	n	critical value
5	1.00	27	0.382	49	0.282	92	0.205
6	0.886	28	0.375	50	0.279	94	0.203
7	0.786	29	0.368	52	0.274	96	0.201
8	0.738	30	0.362	54	0.268	98	0.199
9	0.700	31	0.356	56	0.264	100	0.197
10	0.648	32	0.350	58	0.259		
11	0.618	33	0.345	60	0.255		
12	0.587	34	0.340	62	0.250		
13	0.560	35	0.335	64	0.246		
14	0.538	36	0.330	66	0.243		
15	0.521	37	0.325	68	0.239		
16	0.503	38	0.321	70	0.235		
17	0.485	39	0.317	72	0.232		
18	0.472	40	0.313	74	0.229		
19	0.460	41	0.309	76	0.226		
20	0.447	42	0.305	78	0.221		
21	0.435	43	0.301	80	0.220		
22	0.425	44	0.298	82	0.217		
23	0.415	45	0.294	84	0.215		
24	0.406	46	0.291	86	0.212		
25	0.398	47	0.288	88	0.210		
26	0.390	48	0.285	90	0.207		

1. Adapted from J.H. Zar. *Biostatistical Analysis*. Prentice-Hall, Englewood Cliffs, N.J. 1974.

The t Test and the F Test[1]

Purpose:

The t Test is a parametric test for differences between means of independent samples. The formula for calculating t takes on different forms depending upon whether the two samples being compared have (statistically) equal variances. Step 1 under the procedures is the **F test** which **can establish whether the variances of the samples are significantly different.**

Warnings and precautions:

This test requires :
1. The individual observations be independent of each other;
2. The distribution of observations be continuous; and
3. The observations be normally distributed.
4. If the variances of the samples differ significantly from each other, you calculate t by the formula in step 4.

Notation:

n_1 and n_2 = the sample sizes of sample 1 and sample 2, respectively.

$\overline{X}_1$ and $\overline{X}_2$ = the means of samples 1 and 2, respectively.

Procedure:

1. Determine whether the variances of the two samples in question are significantly different from each other by using the F test described below.

   The $H_0 : s^2_1 = s^2_2$ the $H_1 : s^2_1 \neq s^2_2$

To use the **F test**, you need to calculate a value F and your degrees of freedom. F is calculated by dividing the larger of the two variances by the smaller. If s^2_1 is numerically larger than s^2_2, then $F = \dfrac{s^2_1}{s^2_2}$.

1. The authors thank Donald Y Young, Department of Ecology, Ethology, and Evolution, University of Illinois, Urbana, Illinois, for contributing this test.

The degrees of freedom for the numerator is $n_1 - 1$, and for the denominator is $n_2 - 1$.

The final step in the F test is to compare the F you have calculated with the critical value for F at the appropriate degrees of freedom (d.f.) using the F table on page 91. If $F_{calculated}$ is greater than $F_{critical}$, then you may reject the H_0. If $F_{calculated}$ is less than $F_{critical}$, you have failed to reject the H_0.

Now back to the t test.
2. The H_0 for the t test is that there is no difference between the means of samples 1 and 2. $H_0 : \overline{X}_1 = \overline{X}_2$
3. If the variances are unequal, go to step 4. If the variances are equal, calculate t using the formula below:

$$ t = \frac{\left| \overline{X}_1 - \overline{X}_2 \right|}{\sqrt{\frac{(n_1 - 1)s^2_1 + (n_2 - 1)s^2_2}{n_1 + n_2 - 2}} \sqrt{\frac{1}{n_1} + \frac{1}{n_2}}} $$

4. If the variances are unequal, calculate t as follows:

$$ t = \frac{\left| \overline{X}_1 - \overline{X}_2 \right|}{\sqrt{\frac{s^2_1}{n_1} + \frac{s^2_2}{n_2}}} \qquad \text{with d.f.} = \frac{\left(\frac{s^2_1}{n_1} + \frac{s^2_2}{n_2} \right)^2}{\frac{\left(\frac{s^2_1}{n_1} \right)^2}{n_1 + 1} + \frac{\left(\frac{s^2_2}{n_2} \right)^2}{n_2 + 2}} - 2 $$

5. When you have calculated your value t, and your d.f., refer to the t table on page 90. At the .05 alpha level, and at the

appropriate degrees of freedom, your critical t value is given in the table. If the t you have calculated is *greater than* the critical value for t in the table, you may reject your null hypothesis. Since you have rejected the null that the means were equal, you may accept your H_1 that they are unequal. If $t_{calculated}$ is *less than* $t_{critical}$, you have failed to reject your null hypothesis.

Example:

The next table lists compact automobiles, their corresponding country of origin, and their gasoline mileage (compiled a few years ago from *Road and Track* and *Motor Trend* magazines). You would like to know whether the gas mileage of domestic cars differs significantly from that of imported cars.

The parameters calculated from our domestic car data are:

$$\overline{X}_d = 19.38 \quad s^2_d = 13.98 \quad n_d = 8 \quad (d = domestic)$$

The parameters (miles per gallon) for the imported cars are:

$$\overline{X}_i = 26.92 \quad s^2_i = 16.27 \quad n_i = 12 \ (i = imported)$$

In order to select the appropriate t test formula, we must first determine whether or not our calculated variances are significantly different using the F test.

$$H_0 : s^2_d = s^2_i \quad F_{calc} = \frac{s^2_i}{s^2_d} = \frac{16.27}{13.98} = 1.16$$

At the .05 alpha level, and d. f. of 11 and 7, we can estimate from our F table (page 91) that F_{crit} is approximately 4. (If the F table does not accommodate your particular combination of degrees of freedom, you can interpolate.) Since F_{calc} is less than F_{crit}, we fail to reject the H_0 and conclude that our variances are not different.

Now that we have established that the variances between the two samples are not different, we can use the t formula given in step three to test our H_0 that $\overline{X}_d = \overline{X}_i$ (our H_1 is $\overline{X}_d \neq \overline{X}_i$), and the formula $n_1 + n_d - 2$ to calculate our degrees of freedom. The

Table of data for the t test example.

Make and Model	Nationality	Miles per gallon
AMC Gremlin	USA	21
AMC Hornet	USA	16
Audi Fox	West Germany	27
Austin Martin	Great Britian	24
Buick Apollo	USA	15
Capri 2600	West Germany	24
Chevrolet Vega	USA	26
Chevrolet Nova	USA	18
Datsun B210	Japan	28
Dodge Colt	Japan	28
Fiat 128	Italy	35
Ford Pinto	USA	23
Ford Maverick	USA	19
Ford Mustang II	USA	17
Honda Civic	Japan	30
Mazda RX-3	Japan	18
Opel Rallye 1900	West Germany	26
Toyota Corolla 1600	Japan	27
Subaru GL	Japan	27
Volkswagen Superbug	West Germany	29

degrees of freedom are 12 + 8 - 2, or 18. When we plug in our parameter values at the appropriate places in the formula for calculating t, it looks like this:

$$t = \frac{26.93 - 19.38}{\sqrt{\frac{(12 - 1)(16.27) + (8 - 1)(13.98)}{12 + 8 - 2}} \sqrt{\frac{1}{12} + \frac{1}{8}}}$$

$$t = 4.21$$

From the t table (page 90), we find that our t_{crit} at the .05 alpha level with 18 d. f. is 2.101. Since t_{calc} (4.21) is greater than t_{crit}, we may reject our H_0 in favor of our H_1 and conclude that the imported cars have a significantly different gas mileage from that of the domestic cars.

Try this example:

In studying the effects of fatigue on perception, a researcher decides to examine the amount of time that elapses between the first viewing and the visual "reversal" of a Necker cube (an optical illusion). Five students who had been awake for 20 hours were tested and the time to reversal was recorded in seconds. Eight other students who had only been awake for 3 hours were given the same test and their times recorded.

The data (taken from Terrace and Parker, see bibliography) looked like this:

Group A awake 20 hrs.	Group B awake 3 hrs.
7.2 sec.	4.7 sec.
5.7	6.3
6.4	4.1
5.3	3.9
6.9	5.4
	5.6
	4.4
	4.8

The null hypothesis here would be that the mean time of reversal in seconds was the same for sample A as for sample B.

$H_0 : \overline{X}A = \overline{X}B$

Table of critical values of t for the t Test at the .05 alpha Level.[1]

degrees of freedom	t critical	degrees of freedom	t critical
1	12.706	21	2.080
2	4.303	22	2.074
3	3.182	23	2.069
4	2.776	24	2.064
5	2.571	25	2.060
6	2.447	26	2.056
7	2.365	27	2.052
8	2.306	28	2.048
9	2.262	29	2.045
10	2.228	30	2.042
11	2.201	40	2.021
12	2.179	60	2.000
13	2.160		
14	2.145		
15	2.131		
16	2.120		
17	2.110		
18	2.101		
19	2.093		
20	2.086		

1. Adapted from J.P. Guilford. *Fundamental Statistics in Psychology and Education*. McGraw-Hill Book Company. New York. 1956.

Table of values to be used with the F Test (at the .05 alpha level).[1]

Degrees of freedom in the numerator

Denominator	1	2	3	4	5	6	7	8	9	10	12	15	20	24	30	40	60	120	∞
1	647.8	799.5	864.2	899.6	921.8	937.1	948.2	956.7	963.3	968.6	976.7	984.9	993.1	997.2	1001	1006	1010	1014	1018
2	38.51	39.00	39.17	39.25	39.30	39.33	39.36	39.37	39.39	39.40	39.41	39.43	39.45	39.46	39.46	39.47	39.48	39.49	39.50
3	17.44	16.04	15.44	15.10	14.88	14.73	14.62	14.54	14.47	14.42	14.34	14.25	14.17	14.12	14.08	14.04	13.99	13.95	13.90
4	12.22	10.65	9.98	9.60	9.36	9.20	9.07	8.98	8.90	8.84	8.75	8.66	8.56	8.51	8.46	8.41	8.36	8.31	8.26
5	10.01	8.43	7.76	7.39	7.15	6.98	6.85	6.76	6.68	6.62	6.52	6.43	6.33	6.28	6.23	6.18	6.12	6.07	6.02
6	8.81	7.26	6.60	6.23	5.99	5.82	5.70	5.60	5.52	5.46	5.37	5.27	5.17	5.12	5.07	5.01	4.96	4.90	4.85
7	8.07	6.54	5.89	5.52	5.29	5.12	4.99	4.90	4.82	4.76	4.67	4.57	4.47	4.42	4.36	4.31	4.25	4.20	4.14
8	7.57	6.06	5.42	5.05	4.82	4.65	4.53	4.43	4.36	4.30	4.20	4.10	4.00	3.95	3.89	3.84	3.78	3.73	3.67
9	7.21	5.71	5.08	4.72	4.48	4.32	4.20	4.10	4.03	3.96	3.87	3.77	3.67	3.61	3.56	3.51	3.45	3.39	3.33
10	6.94	5.46	4.83	4.47	4.24	4.07	3.95	3.85	3.78	3.72	3.62	3.52	3.42	3.37	3.31	3.26	3.20	3.14	3.08
11	6.72	5.26	4.63	4.28	4.04	3.88	3.76	3.66	3.59	3.53	3.43	3.33	3.23	3.17	3.12	3.06	3.00	2.94	2.88
12	6.55	5.10	4.47	4.12	3.89	3.73	3.61	3.51	3.44	3.37	3.28	3.18	3.07	3.02	2.96	2.91	2.85	2.79	2.72
13	6.41	4.97	4.35	4.00	3.77	3.60	3.48	3.39	3.31	3.25	3.15	3.05	2.95	2.89	2.84	2.78	2.72	2.66	2.60
14	6.30	4.86	4.24	3.89	3.66	3.50	3.38	3.29	3.21	3.15	3.05	2.95	2.84	2.79	2.73	2.67	2.61	2.55	2.49
15	6.20	4.77	4.15	3.80	3.58	3.41	3.29	3.20	3.12	3.06	2.96	2.86	2.76	2.70	2.64	2.59	2.52	2.46	2.40
16	6.12	4.69	4.08	3.73	3.50	3.34	3.22	3.12	3.05	2.99	2.89	2.79	2.68	2.63	2.57	2.51	2.45	2.38	2.32
17	6.04	4.62	4.01	3.66	3.44	3.28	3.16	3.06	2.98	2.92	2.82	2.72	2.62	2.56	2.50	2.44	2.38	2.32	2.25
18	5.98	4.56	3.95	3.61	3.38	3.22	3.10	3.01	2.93	2.87	2.77	2.67	2.56	2.50	2.44	2.38	2.32	2.26	2.19
19	5.92	4.51	3.90	3.56	3.33	3.17	3.05	2.96	2.88	2.82	2.72	2.62	2.51	2.45	2.39	2.33	2.27	2.20	2.13
20	5.87	4.46	3.86	3.51	3.29	3.13	3.01	2.91	2.84	2.77	2.68	2.57	2.46	2.41	2.35	2.29	2.22	2.16	2.09
21	5.83	4.42	3.82	3.48	3.25	3.09	2.97	2.87	2.80	2.73	2.64	2.53	2.42	2.37	2.31	2.25	2.18	2.11	2.04
22	5.79	4.38	3.78	3.44	3.22	3.05	2.93	2.84	2.76	2.70	2.60	2.50	2.39	2.33	2.27	2.21	2.14	2.08	2.00
23	5.75	4.35	3.75	3.41	3.18	3.02	2.90	2.81	2.73	2.67	2.57	2.47	2.36	2.30	2.24	2.18	2.11	2.04	1.97
24	5.72	4.32	3.72	3.38	3.15	2.99	2.87	2.78	2.70	2.64	2.54	2.44	2.33	2.27	2.21	2.15	2.08	2.01	1.94
25	5.69	4.29	3.69	3.35	3.13	2.97	2.85	2.75	2.68	2.61	2.51	2.41	2.30	2.24	2.18	2.12	2.05	1.98	1.91
26	5.66	4.27	3.67	3.33	3.10	2.94	2.82	2.73	2.65	2.59	2.49	2.39	2.28	2.22	2.16	2.09	2.03	1.95	1.88
27	5.63	4.24	3.65	3.31	3.08	2.92	2.80	2.71	2.63	2.57	2.47	2.36	2.25	2.19	2.13	2.07	2.00	1.93	1.85
28	5.61	4.22	3.63	3.29	3.06	2.90	2.78	2.69	2.61	2.55	2.45	2.34	2.23	2.17	2.11	2.05	1.98	1.91	1.83
29	5.59	4.20	3.61	3.27	3.04	2.88	2.76	2.67	2.59	2.53	2.43	2.32	2.21	2.15	2.09	2.03	1.96	1.89	1.81
30	5.57	4.18	3.59	3.25	3.03	2.87	2.75	2.65	2.57	2.51	2.41	2.31	2.20	2.14	2.07	2.01	1.94	1.87	1.79
40	5.42	4.05	3.46	3.13	2.90	2.74	2.62	2.53	2.45	2.39	2.29	2.18	2.07	2.01	1.94	1.88	1.80	1.72	1.64
60	5.29	3.93	3.34	3.01	2.79	2.63	2.51	2.41	2.33	2.27	2.17	2.06	1.94	1.88	1.82	1.74	1.67	1.58	1.48
120	5.15	3.80	3.23	2.89	2.67	2.52	2.39	2.30	2.22	2.16	2.05	1.94	1.82	1.76	1.69	1.61	1.53	1.43	1.31
∞	5.02	3.69	3.12	2.79	2.57	2.41	2.29	2.19	2.11	2.05	1.94	1.83	1.71	1.64	1.57	1.48	1.39	1.27	1.00

Degrees of freedom in the denominator

1. Adapted from R.G.D. Steel and J.H. Torrie. Principles and Procedures of Statistics. McGraw-Hill Book Co., New York. 1960.

χ^2 (Chi-Square) One-sample Test for Goodness of Fit

Purpose:

This χ^2 test is a test of differences between distributions. It allows one to compare a collection (or frequency distribution) of discrete, nominal data with some theoretical expected distribution to see if they differ significantly. (χ is the Greek letter chi, pronounced kī as in "sky.")

Warnings and precautions:

1. The data must be discrete and nominal.
2. When there are only two categories, no expected value may be less than 5. When there are more than two categories, no more than 20% of the expected values may be less than 5, and no expected value may be less than 1.

Null hypothesis:

The observed frequency distribution (or ratio) is equal to the expected frequency distribution (or ratio).

Formula:

$$\chi^2 = \Sigma \frac{(\text{Obs. - Exp.})^2}{\text{Exp.}}$$

That is, for the different distributions you are comparing, you square the difference between the observed value and the theoretically expected value, and divide this square by the expected value. The figures calculated for each distribution are then summed for the Chi-Square.

If the value calculated for Chi-Square is *equal to* or *greater than* the critical value given in the table on page 95, for your degrees of freedom, you may reject the null hypothesis. The degrees of freedom (d.f.) for the Chi-Square One-Sample Test for Goodness of Fit is equal to the number of categories (or columns, see the example) minus 1.

Example:

You wish to know whether there are more Chevrolets, Fords, or Buicks passing the window of your laboratory during the lunch hour. You count the cars passing from noon to 1:00 P.M and find that there are 56 Chevrolets, 71 Fords and 83 Buicks. For the purposes of your null hypothesis that $56 : 71 : 83 = 1 : 1 : 1$, you need to have a theoretically expected ratio. In this case, the three observed values are summed and divided by three so that you can use your mean as the expected. You would arrange your data in this way:

	Chevrolets	Fords	Buicks	
observed	56	71	83	
expected	70	70	70	
obs. − exp.	−14	1	13	
(obs. − exp.)²	196	1	169	
$\dfrac{\text{(obs. − exp.)}^2}{\text{exp.}}$	2.80	.01	2.41	X^2 = the sum of this line X^2 = 5.22

Since there are three categories, your degrees of freedom = 2. You check the Chi-Square table at your appropriate d.f. and find the critical value is 5.99. Since your calculated Chi-Square value (5.22) is less than the critical value, you fail to reject the null hypothesis. Therefore you can conclude that there is no statistically significant difference among the number of Chevrolets, Fords, and Buicks passing the window during the hour in question.

Try this example:

Let us say that there are five candidates for political office and a sample of 500 people are interviewed by pollsters to see whom they prefer. In testing the data collected, the null hypothesis would be that there is no difference in preference or that the actual preferences recorded do not differ significantly from 100 people in favor of each candidate. In fact, the results of the interviews look like this:

Candidate	Number of People Preferring the Candidate
JOHN	122
DICK	105
HARRY	86
JIMMY	83
JERRY	104

χ^2 Table at the .05 alpha level.[1]

d.f.	critical value
1	3.84
2	5.99
3	7.81
4	9.49
5	11.1
6	12.6
7	14.1
8	15.5
9	16.9
10	18.3
11	19.7
12	21.0
13	22.4
14	23.7
15	25.0
16	26.3
17	27.6
18	28.9
19	30.1
20	31.4

1. Adapted from R.G.D. Steel and J.H. Torrie. *Principles and Procedures of Statistics.* McGraw-Hill Book Co.. N.Y. 1960.

χ^2 Test of Independence Between Two or More Samples

Purpose:

This χ^2 (Chi-Square) test, a test of differences between distributions, is used to test for independence between two or more frequency distributions of nominal data. This test is similar in rationale to the Chi-Square One-Sample Test for Goodness of Fit. The main differences are that data are put in a contingency table and that the calculations of the expected values are different.

Warnings and precautions:

1. Data must be discrete and nominal.
2. When there are only two categories, no expected value may be less than 5. When there are more than two categories, no more than 20% of the expected values may be less than 5, and no expected value may be less than 1.

The null hypothesis:

The distribution of sample A is equal to the distribution of sample B.

Formula:

$$\chi^2 = \Sigma \frac{(\text{Obs.} - \text{Exp.})^2}{\text{Exp.}}$$

That is, for the different distributions you are comparing, you square the difference between the observed value and the expected value, and divide this square by the expected value. The figures thus calculated for each distribution are summed to produce the Chi-Square.

If the value calculated for Chi-Square is *equal to* or *greater than* the critical value given in the table on page 95, for your degrees of freedom, you may reject the null hypothesis. The d.f. for this test = (# columns - 1) (# rows - 1).

Example:

Is the distribution of Fords, Chevys, Buicks and Opels on a given day the same on Henley Street as on Kingston Pike? In this example, the null hypothesis is:

Henley Street   Kingston Pike
#Fords: #Chevys: #Buicks: #Opels = #Fords: #Chevys: #Buicks: #Opels

You make your observations and record your data in a contingency table that looks like this:

		Fords	Chevys	Buicks	Opels	row totals	
Henley St.	obs.	45	72	27	90	234	row 1
	exp.	39.8	69.3	65.3	59.6		
Kingston Pk.	obs.	25	50	88	15	178	row 2
	exp.	30.2	52.7	49.7	45.4		
Column totals		70	122	115	105	412 grand total	

Since the theoretical distribution is not readily obvious, we must calculate the expected value for each cell of the table. You calculate the expected value for each cell by multiplying each cell's row total by its column total and then dividing this product by the grand total. For example, to calculate the upper left expected cell,

$$\frac{234 \times 70}{412} = 39.8$$

The next step of the test is to calculate the

$$\frac{(obs. - exp.)^2}{exp.}$$

for each cell (.68 for the upper left corner). The summation of the above calculations from each cell is the Chi-Square value. The degrees of freedom for this test = (# columns or categories - 1) times (# rows - 1) or, in this case, 3. Chi-Square = 89.66.

When the Chi-Square table is consulted (the table is for the .05 alpha level), at the appropriate d.f., we find that the critical value is 7.81. This value is much smaller than our calculated Chi-Square of 89.66 so we can reject our null hypothesis that the ratios are equal and conclude the two samples (Henley Street and Kingston Pike) are independent.

Try this example:

One researcher examining the status of mentally deficient adults decided to look for a relationship between being married and level of intelligence. In this case a Chi-Square test of independence can be applied to his data to test the null hypothesis there is no relationship between being married and one's level of intelligence.

The data collected looked like this (data are taken from Guilford 1978; see bibliography):

Comparison of men of normal IQ with feebleminded men with respect to marital status			
Marital status	Normal	Feebleminded	Both (sample total)
Married	111	84	195
Unmarried	95	122	217
Total	206	206	412

CHAPTER NINE

Using the Library
Diane Schmidt

Biology Librarian, University of Illinois at Urbana-Champaign

Library basics

Modern university or college libraries are complicated places, and they have only become more complicated with the advent of electronic books and journals. Depending on the size of the institution that you attend, your library may be very large with several branches and millions of volumes, or it may be relatively small and have only one building and tens of thousands of volumes. Large or small, all libraries have the same general organization.

There will be books and journals, CD-ROMs and electronic databases, and almost certainly access to the Web. Most college and university libraries use the Library of Congress classification system, rather than the Dewey Decimal system used at most public school and public libraries, but the underlying principles are the same. Materials on the same subject are grouped together.

To find books on a particular subject, you will need to look in your library's online catalog, which may also be referred to as an OPAC or library catalog, or other similar names. It will probably also have a catchy name such as InfoHawk, THOR, or GrizNet. You can look up the author's name, the book title, or subject words on the online catalog. Most catalogs also allow you to search by keywords, which may be located in any field in the record. Figure 1 shows a fairly typical online catalog record for a book. If you were going to cite this book in a paper, you would need to write down the author, title, edition, publisher, and page numbers. To find the book, you would need to write down the call number. The ISBN, or International Standard Book Number,

Figure 1. Sample book record from online catalog

Author: Mech, L. David.
Title: The wolf: the ecology and behavior of an endangered
    species, by L. David Mech.
Edition: 1st University of Minnesota Press ed.
Published: Minneapolis : University of Minnesota Press,
    1981, c 1970.
Format: xxiii, 384 p. : ill. ; 23 cm.
Subject: Wolves
Notes: Includes indexes.
    Bibliography: p. [364]–374.
ISBN: 0816610266 (pbk.)
Call Number: QL737.C22M4 (Library of Congress
    Classification System)

will not help you find the book in your library. Figure 2 shows a similar online catalog record for a journal. Once again, the ISSN or International Standard Serial Number, will not help you find

Figure 2. Sample journal record from online catalog

Title: The Canadian field-naturalist.
Published: Ottawa, Ottawa Field-Naturalists' Club.
Publication History: v. 33– Apr. 1919–
Frequency: Quarterly
Format: v. ill. 26 cm.
Subject: Natural history—Canada—Periodicals.
Former Title: Ottawa naturalist.
Notes: Nuclear science abstracts 0029-5612
    Biological abstracts 0006-3169
    Selected water resources abstracts 0037-136X
    Bibliography of agriculture 0006-1530
ISSN: 0008-3550
Call Number: 570.5 CAN (*Dewey Decimal Classification
    System*)
Holdings: v. 33 to date, 1919 to date.

the journal. You will probably need the call number, though many libraries shelve their journals by title so the call number is unnecessary.

You will be able to find the locations of journals in an online catalog, but not individual articles. Some online catalogs will allow you to link directly to electronic books and journals once you find their record in the catalog, but some systems do not have this function. Your library may have a separate list of electronic materials.

Most journals and a few books are now available electronically, which means that you can read the article or book on your computer or print a copy. However, your library may not subscribe to the electronic version of a journal or book, since they may be more expensive than the print version or require an extra fee. Another point to consider is electronic journals are generally only available from about 1995 to date. There are rare exceptions, so ask your librarian or check your local online catalog for details about the journal you are seeking.

In fact, the most useful thing you can do if you are confused or having trouble finding information is to ask the librarian for help. Librarians will not do your work for you, but are glad to show you how to do it yourself.

Choosing a research topic

If you are writing a paper, whether it is a scientific article, a lab report, or a term paper, you need to find material about your subject. Many sources of information exist including encyclopedias, journal articles, and Web sites. Encyclopedias and textbooks are a good source for background information or to give you ideas about what topics to study, but you should always back up your conclusions with information from the original research, generally reported in journal articles. Web sites, like encyclopedias, can be good sources of background information but are usually not considered authoritative. Let's take an example of a simple research topic and look at the kinds of information you can find from the various sources mentioned here.

Let's say you want to do a paper on an endangered species. You will need to narrow your topic, since this is a very broad subject. If you look for general background information on the Web or in an encyclopedia such as the *Beacham's Guide to Endangered Species* (mentioned in the next chapter), you will find that, while wolves are rare in the continental United States, they are relatively common in Minnesota.

Perhaps you could do your paper on Minnesota wolves. Even this is a broad topic. Are you interested in predator-prey relations, wolf diseases, the size of wolf populations, wolf behavior, or perhaps something else? Looking for books on your topic may give you other ideas about how to narrow down your topic, but you will need to search for your broad topic. Your library may not have any books on wolves in Minnesota, but it will surely have books on wolves in general.

Finding articles

Once you have picked a reasonably narrow topic, you will need to find more information about it. For scientific areas, this means searching for journal articles. To do this you will need to pick one of the indexes or databases listed in the next chapter. A sample

---

Figure 3. Sample article from index (*Biological Abstracts*)

TI: Tolerance by denning wolves, *Canis lupus*, to human disturbance.

AU: Thiel-Richard-P {a}; Merrill-Samuel; Mech-L-David

AD: {a} Wis. Dep. Nat. Resources, Sandhill Wildlife Area, Box 156, Babcock, WI 54413, USA

SO: Canadian-Field-Naturalist. April–June, 1998; 112 (2) 340–342.

PY: 1998

DT: Article-

IS: 0008-3550

LA: English

AB: Wolves are considered to be intolerant of human activity, especially near dens and pups. In recent years range extensions of the species in the upper Great Lakes region have brought Wolves in closer contact with humans. We report observations of Wolves tolerating human activity in close proximity to dens and rendezvous sites with pups. These include moss harvesting work in the Black River State Forest, Wisconsin; military maneuvers at Camp Ripley Military Reservation, and road construction work in the Superior National Forest in Minnesota.
MC: Behavior-
ST: Canidae-: Carnivora-, Mammalia-, Vertebrata-, Chordata-, Animalia-
OR: *Canis-lupus* [wolf-] (Canidae-)
TN: Animals-; Carnivores-; Chordates-; Mammals-; Nonhuman-Mammals; Nonhuman-Vertebrates; Vertebrates-
GE: Black-River-State-Forest (Wisconsin-, USA-, North-America, Nearctic-region);Camp-Ripley-Military-Reservation (Minnesota-, USA-, North-America, Nearctic-region); Superior-National-Forest (Minnesota-, USA-, North-America, Nearctic-region)

MI: den-proximity; human-disturbance; tolerance-; Note-
AN: 199800427871
UD: 19980803

---

record from an index is shown in Figure 3. To find the particular article shown in the figure, you will need to write down, at a minimum, the information on the Source line (which may be abbreviated as SO or called different things in various databases). Typically, the Source line lists the journal name, volume number, issue number (in parentheses), page numbers, and date. To cite the article in your paper, you will need the author(s) and title of the article as well. The format for a citation is discussed in Chapter Eleven.

A good strategy for searching databases is to think carefully about what terms to use before you start. You may not find any articles if you search for "timber wolves," but "wolves" pulls up many articles. You should always use the scientific name of any organism (in this case, *Canis lupus*) as well as its common name. If you aren't finding very many articles, look at the articles that seem most useful. There should be a field labeled "keywords" or "subject terms" or something similar. These are the terms the author or someone else has used to describe the article. If you search using the same words, you may have better results.

The articles that you find will vary depending on the database you use. For instance, if you search "wolves and Minnesota" in *Biological Abstracts*, you will find articles on topics such as the effect of wolves on the white-tail deer harvest, the tolerance of denning wolves to human disturbance, the incidence of coccidiosis in wolf populations, and the like. What if you want to find articles that deal with the interaction between wolves and livestock? If you can't find useful articles in *Biological Abstracts*, perhaps you should try AGRICOLA, an agricultural database. Here you will find articles about Lyme disease in wolves, wolf control efforts, and the management of wolf-livestock conflicts.

If you are more interested in articles about the controversies relating to the conservation status of wolves in Minnesota, you might be better off with a more general index such as *Biology Digest* or even the *Reader's Guide to Periodical Literature*. You will find articles on wolf resettlement, hunting, and Wolf-Rayet stars in the *Reader's Guide*. The articles about Wolf-Rayet stars illustrate one problem with searching general indexes—words mean different things in different fields.

Most article databases will only give you information about the articles: the author, title of the article, title of the journal, date, volume, page numbers and perhaps the abstract. To find the actual article, you will need to take the information from the database and find the article itself in the journal. That may involve searching your online catalog to see if your library subscribes to the journal, or looking up the journal in a list of electronic journals.

No library subscribes to all the journals indexed in a database, so chances are at least some of the articles you are interested in are not available. One common mistake students make is to search their online catalog for the author or title of the article, not the title of the journal. Some versions of article databases will allow you to click on a link and go directly to the full text of an article, but this only works if your library subscribes to that journal. PubMed, the free medical database mentioned in the next chapter, is a good example of one of these one-click databases.

Searching the Web

If all this sounds very complicated and time-consuming, it can be. There is a lot of temptation to do your research the easy way, by doing a nice, quick Web search. Sometimes you can find valuable information on the Web, but you need to take a hard look at the results you get. To go back to our "wolves in Minnesota" search, in March 2001 a quick search on the two terms in Google pulled up about 200,000 sites. Among the first few sites listed were the Minnesota Department of Natural Resources (DNR) site, the International Wolf Center's site, several wolf-lover's home pages, wolf books and other gifts, the Minnesota Timberwolves basketball team's site, CNN and ABC news reports about wolves, a site sponsored by People Against Wolves, and the Wolf Lake, Minnesota city directory.

Even when you eliminate the non-*Canis lupus* sites you are left with a lot of dubious material. Depending on which site you are looking at, wolves may be noble beasts, ravenous monsters, or somewhere in between. Pro-wolf sites are more common than anti-wolf sites, at least in the first few sites retrieved. But this doesn't necessarily mean more people are in favor of having wolves in their back yards; perhaps pro-wolf people have better access to computers than anti-wolf people do.

When contrasting Web sites describe wolves as either harmless or dangerous, who are you going to believe? This is when the prudent Web searcher takes a close look at the data provided.

Are sources listed, and how reliable are they? Hearsay evidence is weaker than carefully tested data. Do the people sponsoring the site have an axe to grind? What do sites such as the Minnesota DNR or the US Fish and Wildlife Service say about the subject? Data published in scientific journals and on some Web sites receive scrutiny from the peer review process (described in the next chapter), but most of the Web pages do not have that stamp of approval. However, some authoritative information is available from the Web. For instance, the US Fish and Wildlife Service's Endangered Species site will give you the regulatory profiles for endangered species, available in numerous print publications as well. The Web site information may be more up-to-date than your library's books, so this is one case where the Web can really help you find information.

In some fields, the Web is the primary source of data and provides reliable information in those areas. Molecular biologists, for instance, rely on Web databases such as GenBank and the DNA DataBank of Japan (DDBJ). Many taxonomists provide authoritative information about organisms on their institution's Web sites as well. Scientific associations may also provide some useful information for students on their Web sites, and the Web directories listed in the next chapter link to carefully chosen sites. Still, you should usually use the Web for background research, not as your main source of information and most items that you cite in papers should be regular journal articles, whether you found them in an electronic journal or in a paper journal.

Recording your findings

As you work, be sure to record all of the books, articles, and Web sites you have consulted. A time-honored method is to use 3" x 5" index cards for each reference. In addition to the factual information you need for your paper, record all the bibliographic information for the "literature cited" section. Be sure the information you write down includes the correct name of the author(s) and the year of publication. For articles, you will also need the title of the article and the journal, the volume and issue numbers, and the page numbers of the article. For books, you will need the publisher name and location, the edition (if any), and the number of pages in the book. If you are quoting from an article or book, make sure you also record the page number where the quotation is found.

If you copy an entire article and highlight relevant passages rather than using 3" x 5" cards, you may not need to record all this information. Most articles include all the proper citation information at the top or bottom of the first page of the article, but, depending on your chosen citation format, some of this information may be missing. Check carefully before you put the journal volume back. It will save you a lot of trouble later.

Citing Web sites is a more complicated issue than citing books or articles. You should copy down the Web address or URL, but also look at the site to see if you can find the same kind of information that you would need to cite a book or article. Try to identify an author, whether a person or an institution. You may also be able to find a date and a location. It may not be possible to find all of the information you would usually use; but, at the very least, your citation should include the title of the Web site, the URL and the date you looked at it. The *Columbia Guide to Online Style* mentioned in the next chapter provides some good examples of how to cite Web pages. The *Tree of Life* Web site listed in Chapter Ten is in a format based on the *Columbia Guide*.

No matter what source you use to find information, make sure your citations are correct and complete. Citations from article databases are usually accurate, but even they occasionally make mistakes. Citations taken from other papers are often incorrect, and you do not want to perpetuate mistakes that reflect poorly on your research. Likewise, you should always examine the papers you cite, even if it is tempting to take information from just the abstract or title. Papers that *look* relevant may not be, and this kind of shoddy research is even worse than inaccurate citations. You may even get a failing grade if you try to cut corners.

CHAPTER TEN

An Introduction to Biological Literature

Diane Schmidt

Biology Librarian, University of Illinois at Urbana-Champaign

The published literature of biology is vast and varied. It includes publications such as textbooks, encyclopedias, book series, dissertations, patents, journal articles, indexes, Web sites, and much more. There is a definite hierarchy among the kinds of publications. Journal articles and other forums that publish original research are considered primary literature and are the most prestigious. Reference sources that repackage the primary literature, such as encyclopedias, handbooks, indexes, and other sources, are called the secondary literature. The tertiary literature includes guides to the literature (such as this chapter) that point users to useful secondary sources.

The main venue for publishing biological research is the journal article. The first scientific journal was the *Philosophical Transactions of the Royal Society of London*, which was first published in 1665 and is still around today. The French *Journal des Scavans* was published the same year, and, ever since, the number of journals has grown rapidly. No one is quite sure how many scientific journals are being published around the world, but there are probably at least 30,000 of them. Not all are important, but even so, the number of journals a scientist might use in a lifetime is astonishing.

One major feature of the current model of scientific publishing is peer review. When a scientist sends an article to an editor for publication in a journal, the editor sends the manuscript to two or more other scientists who are knowledgeable in that field. The reviewers thoroughly examine the article to see if it is accurate, properly written, and contains new information. Articles not passed by reviewers may be sent back to the author for changes or rejected outright. Other publications such as conference papers, technical reports, patents, and Web sites are usually viewed as inferior because they have not received peer review.

Some other scientific fields are not as tied to peer review as biology. For instance, physicists have made heavy use of Web preprint servers. Before an article is accepted by a physics journal, a version of the article may be made available on a Web site. This allows access to the research very quickly, and the physicists are comfortable with the lack of peer review. They say they can tell who the "best" authors are, and don't care about the rest. Biologists, especially biomedical researchers, are generally against preprints since they are concerned about non-specialists finding inaccurate studies that might be dangerous or misleading.

Some of the major secondary resources that students may find useful are listed below. This is only a very selected list of common books and indexes that are available in most libraries, and you should follow the steps outlined in the previous chapter to find other relevant materials.

Dictionaries and Encyclopedias

Many excellent dictionaries and encyclopedias are available for biologists. Several major publishers have developed specialized encyclopedias for areas such as microbiology, immunology, bioethics and biotechnology. These are a very good place to start your research.

Beacham's Guide to the Endangered Species of North America. Walton Beacham, Frank V Castronova, and Suzanne Sessine, editors. Detroit (MI): Gale Group; 2001. 7v. ISBN 0787650285 (set).
Covers more than 800 endangered species of plants and animals. Most entries are illustrated with a color photo and include descriptions, ecology, and the cause of the threat. Updates *The Official World Wildlife Fund Guide to Endangered Species of North America.*

Encyclopedia of Bioethics. Rev. ed. Warren T Reich, editor. New York: Macmillan Reference; 1994. 5 v. ISBN 0028973550 (set).
This encyclopedia contains 460 authoritative, original articles, covering ethical and moral dimensions of scientific and technological innovations. Each article includes a list of references.

Encyclopedia of Human Biology. Renato Dulbecco, editor. San Diego (CA): Academic; 1997. 9 v. ISBN 0122269705 (set).
Over 673 articles, each averaging about 10 pages, provide authoritative, up-to-date information on all aspects of human biology.

Encyclopedia of Life Sciences. New York: Nature Publishing Group; 2001. 20 v. ISBN 1561592749.
This major new encyclopedia features more than 4,000 articles by well-known authorities. There is a Web version as well called ELS Online which your library may also subscribe to.

The Facts on File Dictionary of Biology. 3rd ed. Robert Hine, editor. New York: Facts On File; 1999. 361p. ISBN 0816039070, 0816039089 (pa).
Contains 3,300 entries defining the most commonly used biological terms.

Grzimek's Animal Life Encyclopedia. Bernhard Grzimek, editor. New York: Van Nostrand Reinhold Co.; 1972–75. 13 v.
A comprehensive encyclopedia covering animals from protozoa to mammals. This is a grand old classic, and is available in almost every library. The section on mammals was revised in 1990 and is available as *Grzimek's Encyclopedia of Mammals.*

Dictionary of Gardening, Anthony Huxley, editor. New York: Stockton Press; 1992. 4 v. ISBN 1561590010.
This set includes over 50,000 plant entries with brief description and distribution information for important species. All plant genera cultivated anywhere in the world are covered here, including those in botanical gardens. It is one of the most comprehensive sources for basic information about almost any genus of plants.

Henderson's Dictionary of Biological Terms. 12th ed. Isabella Ferguson Henderson. Eleanor Lawrence, editor. London: Prentice Hall; 2000. 719 p. ISBN 0582414989.
Provides definitions for more than 23,000 terms in biology, botany, zoology, anatomy, cytology, genetics, embryology, and physiology.

McGraw-Hill Dictionary of Bioscience. Sybil P Parker, editor. New York: McGraw-Hill; 1996. 448 p. ISBN 0070524300

(paper).
This dictionary defines nearly 16,000 major terms in the biosciences taken from the *McGraw-Hill Dictionary of Scientific and Technical Terms*.

McGraw-Hill Encyclopedia of Science and Technology. 8th ed. Sybil P Parker, Editor-in-Chief. New York: McGraw-Hill; 1997. 20 v. ISBN 0079115047 (set).
This is one of the most commonly used science encyclopedias. Your library may subscribe to the Web version, which is known as *AccessScience@McGraw-Hill*. It is also available on CD-ROM as the *Multimedia Encyclopedia of Science and Technology*.

The Merck Index: An Encyclopedia of Chemicals, Drugs, and Biologicals. 12th ed. Rahway (NJ): Merck; 1996. 1 v. ISBN 0911910123.
This encyclopedia covers more than 10,000 significant drugs and chemicals.

Plant Sciences for Students. Richard Robinson, editor. New York: Macmillan Reference; 2000. 4 v. ISBN 002865434X (set).
This encyclopedia is designed for high school students and undergraduates, covering topics such as careers in plant sciences and biographies of famous botanists.

Handbooks

Handbooks usually consist of summarized information in the form of charts, graphs, tables and the like and are designed to be one-volume reference books. Multivolume works sometimes call themselves handbooks as well, but in general a handbook is a useful compilation of data too tedious to search out in the original sources.

CRC Handbook of Chemistry and Physics. 81st ed. Boca Raton (FL): CRC Press; 2000. 2,556p. ISBN 0849304814, 849308798 (CD-ROM).
This is the standard source for chemical, physical, and engineering data, including: mathematical tables, elements, inorganic compounds, general chemical tables, physical constants, and so on. Your library may subscribe to the Web version as well.

The Diversity of Living Organisms. RSK Barnes, editor. Malden (MA): Blackwell Science; 1998. 345 p. ISBN 0632049170.
This is an illustrated guide to all types of living organisms from the single-celled prokaryotes to multicellular organisms. It examines organisms only to the level of the class, but covers even the most obscure groups.

Flora of North America, North of Mexico. Flora of North America Editorial Committee and Nancy R Morin, editors. New York: Oxford University Press; 1993– .
This projected 30-volume set will provide information on all of the more than 20,000 plant species that grow from the Florida Keys to the Aleutian Islands. The set will be updated by a computer database for taxonomic information housed at the Missouri Botanical Garden in St. Louis, Missouri. See their Web site at http://hua.huh.harvard.edu/FNA/

Information Sources in the Life Sciences. 4th rev. ed. HV Wyatt, editor. New Providence (NJ): Bowker/Saur; 1997. (Guides to Information Sources). 264 p. ISBN 1857390709.
This guide contains chapters discussing various types of resources such as newsletters and databases. In addition, several subject chapters discuss the literature of a specific field. It has a British focus.

The Tree of Life: A Multi-Authored Distributed Internet Project Containing Information About Phylogeny and Biodiversity. DR Maddison and WP Maddison. 1998. http://phylogeny.arizona.edu/tree/phylogeny.html
The project is intended to provide a means for finding information on all taxa of living organisms. The project is unfinished, but provides a good starting point for exploring the natural world.

Using the Biological Literature: A Practical Guide. 3rd ed. Diane Schmidt and Elisabeth B Davis. New York: Marcel Dekker; 2001. In press.
This guide covers the literature of all of the biological sciences, including major Web resources such as freely accessible databases and taxonomy sites.

Scientific Style and Format: The CBE Manual for Authors, Editors, and Publishers. 6th ed. New York: Cambridge University Press; 1994. 825 p. ISBN 0521471540.

This manual covers publication style guidelines (i.e., how publications should be styled and formatted) for all of the scientific disciplines. There are specific guidelines for scientific conventions in such areas as chemical names, cells and genes, and astronomical objects. There are also sections on formatting for books, journals, and other publication types, and information on the publishing process.

Synopsis and Classification of Living Organisms. Sybil P Parker, Editor-in-Chief. New York: McGraw-Hill; 1982. 2 v. ISBN 0070790310 (set).
The systematic positions and affinities of all living organisms are presented in concise articles for all taxa down to the family level.

The Columbia Guide to Online Style. Janice R Walker and Todd W Taylor. New York: Columbia University Press; 1998. 218 p. ISBN 0231107889, 0231107897 (paper).
A very highly regarded guide to citing e-mail and discussion group messages, Web pages, and numerous other electronic sources. Much of the information in the guide is also freely available on the Web at http://www.columbia.edu/cu/cup/cgos/idx_basic.html

Web Directories

Web directories can be useful ways to locate valuable Web sites, sometimes better than using Web search engines such as Google or HotBot. Many directories were created by experts. In addition to the directories listed below, major associations such as the Ecological Society of America often have useful collections of links their members have suggested.

INFOMINE: Biological, Agricultural, and Medical Sciences. Riverside (CA): University of California-Riverside. http://infomine.ucr.edu
INFOMINE is a scholarly Internet resource collection that provides links to resources such as online databases, textbooks, newsgroups, and subject guides.

UniGuide: Academic Guide to the Internet. Aldea Communications; 1998– . http://www.aldea.com/guides/ag/attframes2.html
This directory is arranged by broad categories and, like INFOMINE, it

concentrates on college level resources. Each item is reviewed by librarians and content experts. Formerly called the InterNIC Academic Guide to the Internet.

WWW Virtual Library: Bioscience. WWW Virtual Library; 1994– .
http://www.vlib.org/Biosciences.html
This venerable directory has been around since the beginning of the Web and is supported by a number of societies. It is arranged by subject and is a good starting point for almost any topic.

Reviews of the Literature

Reviews of the literature are state-of-the-art reports of research and techniques covering a specific time period for a specific subject. They are useful for gaining background information on a subject or for finding out what the current research directions are. Many journals publish one or two review articles in each issue, but many journals publish only reviews. Review serials are easy to spot because their titles often begin with words such as *Advances in...*, *Annual Review of...*, *Critical Reviews in...*, *Current Opinion in...*, *Current Topics in...*, *Progress in...*, *Trends in...*, or *Yearbook of....* Reviews may provide background information, may indicate classic papers, and may point out directions for future research. Whatever their aim, a lengthy bibliography of citations to the literature is included in the review. Reviews to the literature are indexed in most of the general abstracts and indexes listed below. Generally, you can find them by searching for your topic and adding the word "reviews."

Abstracts and Indexes (also known as Article Databases)

The major abstracts and indexes used to find biological journal articles are listed here. Most of these indexes include abstracts for the articles that they cover, which can be helpful in determining whether a particular article is worth pursuing. All of the indexes listed here are available in multiple formats, so your library may subscribe to them in print, on CD-ROM, or as a Web-searchable database. You will need to check your library's online catalog to find out which indexes are available for which time periods, and in which format(s). **AGRICOLA** and **PubMed** are available for free to anyone, however, so you can search them using any computer that is connected to the Web.

AGRICOLA (**AGRIC**ultural On**L**ine Access)
This database is produced by the National Agriculture Library and covers the worldwide literature of agriculture including journal articles, monographs, government documents, technical reports and proceedings. This database is valuable for life sciences students interested in plants or animals of economic importance. Available at no charge for anyone to search on the Web at http://www.nal.usda.gov/ag98/

Biological Abstracts. Philadelphia: BIOSIS; v. 1– , 1926– . ISSN 0006-3169.
The most comprehensive biological abstracting service in the world. More than 9,000 journals reporting original research are scanned. The index covers all subjects in biology and biomedicine and is always a good place to start researching a topic in almost any area. *Biological Abstracts* covers only articles reporting original research, and does not include items such as book reviews or letters to the editor. The electronic version of *Biological Abstracts* is often known as BIOSIS Previews.

Biological Abstracts/RRM (Reports, Reviews, Meetings). Philadelphia: BIOSIS; v. 18– , 1980– . ISSN 0192-6985.
Companion to *Biological Abstracts.* Covers non-article material such as editorials, reports, symposia, books, chapters, review journals, and so on.

Biological and Agricultural Index. New York: Wilson; v. 1– , 1916–18– . ISSN 0006-3177.
Appropriate for beginning students and the public. Covers 225 core journals, including book reviews, letters to the editor and other non-article items.

Biological Sciences Collection. Bethesda (MD): Cambridge Scientific Abstracts; 1980– .
This computerized index is the combination of 21 abstracting journals from Cambridge Scientific Abstracts.

Biology Digest. Medford (NJ): Plexus; v. 1– , 1974– . ISSN 0095-2958.

This digest covers about 300 biological journals. The reading level is appropriate for undergraduates and the general public. Unlike *Biological and Agricultural Index*, *Biology Digest* includes abstracts, which are arranged by subject.

Chemical Abstracts. Columbus (OH): Chemical Abstracts Service; v. 1– , 1907– . ISSN 0009-2258.

Covers more than 8,000 journals, plus patents, conference proceedings, reports, and monographs. Essential source for biological topics with chemical facets.

Current Contents/Agriculture, Biology, and Environmental Sciences (CC/ABES). Philadelphia: Institute for Scientific Information; v. 1– , 1970– . ISSN 0011-3379.

Compilation of tables of contents of over 900 major journals as well as major book series. Subjects covered in *CC/ABES* include agriculture, botany, entomology, ecology, mycology, ornithology, veterinary medicine, and wildlife management. The most current listing of journal contents available; much more timely than most indexes.

Current Contents/Life Sciences. Philadelphia: Institute for Scientific Information; v. 1– , 1958– . ISSN 0011-3409.

Companion to *CC/ABES*, above, covering topics such as biochemistry, biophysics, genetics, immunology, microbiology, neurosciences, and pharmacology. Available in 600 and 1,200 title versions.

PubMed. Bethesda (MD):National Library of Medicine;1966– .

PubMed is the free version of the MEDLINE database and offers a number of services aimed at the biological research community. In addition to indexing journal articles, *PubMed* provides links to articles from more than 700 full text journals and to the molecular biology databases of DNA and protein sequences and 3-D structure data that have been developed by NCBI. It is available for free at the National Center for Biotechnology Information's site at http://www.ncbi.nlm.nih.gov/entrez/query.fcgi

Science Citation Index. Philadelphia: Institute for Scientific Information; v. 1– , 1961– . ISSN 0036-827X.

Covers about 4,500 journals in all the sciences. The citation index groups all articles that have referenced the same earlier work so you can find out who cited a particular paper. This provides a different way of finding articles on a topic.

Zoological Record. Philadelphia: BIOSIS; v. 1– , 1864– . ISSN 0144-3607.

The most comprehensive zoological index in the world, this index includes books, proceedings, and over 6,500 periodicals. The most comprehensive source for taxonomic information.

CHAPTER ELEVEN
How to Write a Scientific Paper

The organization of a research paper reflects the basic pattern of research design. Scientific papers follow a rigid format that is extremely helpful to writer and reader. Although the highly structured format might be new to you, you will probably find that writing a scientific paper is easier than writing a paper for a humanities or an English course. Cleverness, beauty, originality, and style (although extremely important in the design of experiments) are not required in scientific writing. What is required is a clear, logical, orderly presentation of your question, how you planned to answer it, what your results were, and what you concluded. As with any writing, good grammar and precise wording are crucial to effective communication.

The scientific paper has the following elements: Title, Abstract (or sometimes a final summary instead), Introduction, Materials and Methods, Results, Discussion, Acknowledgments and Literature Cited. These words are used to head the sections of your paper, are centered and are followed by the text for that section. Normally you do not begin a new page for a new section unless the preceding section completely filled the page. Illustrative tables and figures may be inserted where appropriate or placed at the end of the paper. These items must be fully labeled so they would be understood by someone who had not yet read the paper (for example: not just "Figure 1" but with an explanation of what is being shown; in a graph the axes must be named and the units of measurement given). Data or experimental results presented in a graph or table must also be summarized verbally in the text so someone who has not seen the table can understand the text.

In this chapter, we describe each section of a scientific report in the order in which it appears in the final version. However, this is not the order in which you should write the paper. In general, we recommend starting with the Materials and Methods, which should already be mostly written from your proposal. You should then work on the Results section, including tables and figures. This way you know exactly what you can, and more importantly, what you cannot conclude from your data. The Discussion should be tackled next, and then the Introduction. If you do it in this order, you are better able to direct your readers (in the Introduction) to the precise question you actually answer (in the Discussion). The Abstract summarizes the entire paper and therefore should be written last.

We will describe and illustrate each section of a scientific paper using examples from actual published material and also from unedited papers written by students taking an introductory biology course taught in the investigatory laboratory mode.

I. The Title

The title of a scientific paper should tell the reader what kind of work is being reported. If possible, it should reveal the organism studied, the particular aspect of the system studied, and the variable(s) manipulated. A common student failing is to give a paper an uninformative title such as "Ecology Experiment." Titles should be simple, direct, and informative, using the fewest possible words to convey your meaning. The title is a label; it is not a complete sentence.

For published articles, the title will be read more than any other part of the paper. For example, most people subscribing to academic journals read the titles of all the papers, even though they read the full text of only a few articles. Thus, when you write your title, think about the fact that, unless it convinces them to keep reading, this will be all that most people read. Here are some sample titles:

FRECKLES AND HAIR COLOR: A SEARCH FOR SEX-LINKAGE

WEB-SITE SELECTION IN THE DESERT SPIDER *AGELENOPSIS APERTA*

A NEW SPECIES OF LIZARD OF THE GENUS *AMEIVA*
(TEIIDAE) FROM THE PACIFIC LOWLANDS OF
COLOMBIA

COMPARATIVE ASPECTS OF INVERTEBRATE
EPITHELIAL TRANSPORT

SEASONAL EFFECTS OF DEHYDRATION ON UREA
PRODUCTION IN THE FROG *RANA PIPIENS*

II. The Abstract

The Abstract is a one or two paragraph condensation of the entire article, which briefly describes the results and significance of the study. It helps the reader decide whether the material in the paper will be worth reading. Abstracts are often published by themselves in literature databases such as *Biological Abstracts* (see Chapter Ten). Reducing a long paper to a few paragraphs is an art and naturally requires practice. In general, devote one or two sentences of your abstract to each section of the paper. Here are some sample abstracts—two from published reports and two from student papers.

EVIDENCE FOR A GRADIENT OF A MORPHOGENETIC
SUBSTANCE IN THE DEVELOPING LIMB
Jeffrey A. MacCabe and Brenda W. Parker, 1976
Developmental Biology 54:297–303

Abstract
A polarizing activity in the developing vertebrate limb bud has been implicated in the control of the morphogenesis of its anteroposterior axis. The results of an earlier experiment suggest that there is a gradient of a similar morphogenetic activity in the 4-day chick wing, with a high level of activity along the posterior border, no detectable activity at the anterior border, and an intermediate level of activity in the center, i.e., about halfway between the anterior and posterior borders. The experiments reported here show that this activity disappears from the center

of the limb after placing an impermeable barrier posterior to the center, but not after placing the barrier anterior to the center. A porous barrier placed posterior to the center does not alter the activity in the center of the limb. These results lead us to suggest that the gradient of morphogenetic activity in the chick wing is the result of the movement (possibly by diffusion) of a factor(s) from a source in the posterior region of the wing.

PREY PREFERENCE AND HUNTING HABITAT SELECTION IN THE BARN OWL

Stephen J. Fast and Harrison W. Ambrose III,1976
American Midland Naturalist 96 (2):503–507

Abstract

A series of experimental situations permitting a barn owl (*Tyto alba*) various combinations of choices between woods-like and field-like habitats, and between *Peromyscus leucopus* and *Microtus pennsylvanicus* as prey items, demonstrated that the owl has statistically significant preferences for *Microtus* and the field hunting habitat.

COMPARATIVE CHEMICAL ANALYSIS OF WATER FROM TWO PONDS OF DIFFERENT AGES

Michael Brown and Judy Conley
(students–University of Illinois)

Abstract

Two man-made lakes differing in age by about five years were compared on a chemical basis. Spectrophotometric assays were performed for nitrate, sulfate, phosphorus, lead, and anionic surfactants. A significant difference was found in the mean sulfate and phosphorus concentrations. Environmental implications of this finding and predictions about the aquatic flora are discussed.

THE EFFECT OF VARIOUS LIGHT CONDITIONS ON MATING OF *DROSOPHILA MELANOGASTER*

James F. Palma and Carole Cunningham
(students–University of Illinois)

Abstract

The purpose of our experiment was to determine whether light conditions have any effect on the mating of fruit flies. We obtained virgin fruit flies and placed males and females in agar-filled test tubes. We placed the test tubes under bright, normal and dim light conditions and observed the flies to see if mating took place in any or all of these conditions. We found that flies had no contact in the normal and darkened conditions, while in the bright light, all the flies mated within two hours.

III. The Introduction

The introduction should present the question being asked and place it in the context of what is already known about the topic. Think of the introduction as a funnel. Start broad and gradually focus, always ending the introduction with a clear statement of the research question being addressed. Introductions are written in the present tense. You should define specialized terminology (i.e., jargon) and any abbreviations you will use throughout the rest of the report. Be concise; do not include extraneous information.

Starting broadly places your study within a context, and helps you to describe the relevance of your question (i.e., why is this study interesting?). If your paper is for a class project, then starting broadly also convinces your instructor that you understand the significance of your study. Include background information on your chosen topic that suggests why your question is of interest. For example, if your chosen topic was ant scent trails, you may give background information on chemical communication in insects. When you provide background material, it is very important to give appropriate credit to related studies by citing this work. The format for citing other works will be discussed in Chapter Thirteen.

After a first broad paragraph, you now must channel the introduction towards your specific question. The second paragraph often introduces the specific topic of study, including more background information on what is known about this topic (again, citing all relevant studies). Introductions may also provide relevant information that the non-specialist (i.e., researchers not familiar with the topic or the organism) must learn before they can appreciate the experimental design. To continue the ant example, you might now focus on chemical communication in ants specifically, and the known functions of ant scent trails. This immediate context, if carefully worded, raises the question of how long scent trails last—leading directly to your research question.

Finally, and most importantly, **the last paragraph of your introduction must provide an explicit statement of your research question.** For example, you may start the paragraph with: "In this study, we..." or "The objective of this study was to...." Often, the general approach or methodology of the study is briefly mentioned here, as well as the main result and conclusion.

Some sample introductions from published material follow.

THE ROLE OF STORED GLYCOGEN DURING LONG-TERM TEMPERATURE ACCLIMATION IN THE FRESHWATER CRAYFISH, *ORCONECTES VIRILIS*
Arthur M. Jungreis, 1968
Comparative Biochemistry and Physiology 24:1–6

Introduction
In Crustacea, glycogen metabolism has been associated with chitin synthesis (Renaud, 1949; Passano, 1960). However, the role played by glycogen in intermediary metabolism under normal conditions or during starvation is not clear. According to Renaud, crustacean intermediary metabolism centers around glycogen and fatty acids, while according to Scheer and co-workers (Scheer & Scheer, 1951; Kincaid & Scheer, 1952; Scheer et al., 1952; Neiland & Scheer, 1953) the primary energy source is protein and not carbohydrate and fat. When ^{14}C-labeled glucose was injected into the spiny lobster *Panulirus*, it appeared almost exclusively as labeled glycogen rather than labeled CO_2 (Scheer & Scheer, 1951). Furthermore, during an artificial period of

starvation, the total glycogen content in *Panulirus* did not decrease (Scheer & Scheer, 1951).

To determine the role of glycogen as a metabolic source of energy during acclimation in long-term starved crayfish *Orconectes virilis*, glycogen content was analyzed in a variety of tissues after 45 days of acclimation.

THERMAL BALANCE AND PREY AVAILABILITY: BASES FOR A MODEL RELATING WEB-SITE CHARACTERISTICS TO SPIDER REPRODUCTIVE SUCCESS
Susan E. Riechert and C. Richard Tracy, 1975
Ecology 56 (2):265–284

Introduction

A definite pattern observed in the local distribution of animals can signify the underlying adaptations of individuals to their physical environment. Within a diverse group, those individuals that find suitable locations not only increase their chances of survival to maturity, but also are likely to contribute the greatest number of offspring to the next generation, i.e., exhibit greater fitness (Williams, 1966; Pianka, 1974). Therefore, the genotypes of individuals demonstrating greater habitat discrimination will predominate. This adaptation by the majority of the individuals to selection of favorable habitat is reflected in the local pattern of the population.

Agelenopsis aperta (Gertsch) is a member of the funnel web-building spider family, Agelenidae (Araneae). The web consists of a flat sheet with an attached funnel extending into some feature of the surrounding habitat. Occasionally a scaffolding is present. The sheet has no adhesive properties and serves merely as an extension of the spider's legs. *Agelenopsis* carries out much of its activity within a sheltered environment, coming out of the funnel only as long as required for securing prey and repairing the web. Study of the local distribution of this spider has demonstrated the presence of patterns related to three functions:

124

social, reproductive, and vectorial (Riechert et al., 1973; Riechert, 1974b). A vectorial pattern is influenced by factors of the external environment (e.g., gradients of temperature and humidity, Hutchinson, 1970) and in this context is observed in the association of spiders with specific habitat features. For the funnel web spider, at least, this association does not result from differential survival of randomly dispersed individuals, but rather reflects the active selection of specific web-site characteristics by the spiders (Riechert, 1973). We postulate that individuals, in selecting certain habitat characteristics, are more fit than those showing less, or erroneous discrimination. In this paper, we assess the relationship between the presence of various web-site characteristics and the reproductive success of individuals at web sites offering these characteristics.

DISTINCTIVE FLAVORS INFLUENCE MIXING OF
NUTRITIONALLY IDENTICAL FOODS BY
GRASSHOPPERS
Kerry L. Bright and Elizabeth A. Bernays, 1991
Chemical Senses 16 (4):329–336

Introduction
Polyphagous insects may switch between foods in relation to nutritional needs or for other reasons. The ability of grasshoppers to learn to select foods related to dietary need is well established (Simpson and Simpson, 1990; Simpson and White, 1990; Champagne and Bernays, 1991). The initiation of a change in dietary preference may first be seen as a rejection of the most recently eaten food, either following contact, or early in a meal before normal repletion (Lee and Bernays, 1988). It may also be indicated by direct orientation to food items that are perceptually different (Simpson and White, 1990). Although not shown in grasshoppers, it is also possible that individuals may simply move further away than usual from a recent feeding site, increasing the likelihood of encountering something different (Cohen et al., 1988). Even when foods are nutritionally sufficient, polyphagous species of animals are known to include other items in the overall

diet (Geissler and Rollo, 1988), and these additional items do not necessarily provide any improvement in overall nutrient intake or nutrient balance (Johnson and Collier, 1987). What induces an animal to remain with a good diet or to change to feeding on an alternative one that is the same or inferior?

This study is an investigation of food selection by the grasshopper *Schistocerca americana* Drury, using artificial diets that support good growth, but that vary only in their flavors, in an attempt to separate flavor and nutritional value as bases for dietary selection behavior.

IV. Materials and Methods

This section of the paper should describe the materials and procedures you used in sufficient detail that others could repeat the research. Do not just list the materials you used; describe them as they are relevant to the methods you used. Specific, published techniques can be cited without being described in detail. The method of approach to the problem should be narrated in the **past tense**, describing what was done, not describing how the reader should proceed. Your descriptions of your experiments need to be explicit and thorough; however, take care not to give unnecessary or irrelevant detail. Ask yourself, are these details essential to the success of the experiment? If the answer is no, do not include them.

Methods sections are frequently subdivided, with subheadings for each part of the study. This can make the material much more accessible and useful to readers by breaking it into sections that are easier to read. Often, the first section describes the natural history of the study organism. Common subheadings include: Experimental Design, Statistical Analyses. If several experiments were conducted, subheadings for each would be appropriate. Diagrams of unusual experimental apparatus or study areas (i.e., maps) are often helpful.

Students sometimes make the mistake of describing the results of experiments at the same time they describe the experiments themselves. It is very important not to present results in the Materials and Methods section because most readers will skip or quickly skim this section, and move directly to the Results section. If any of your results are in the Methods section, they will be overlooked.

Since this is often a very long portion of the paper, in some of the following examples, excerpts will be given rather than entire sections.

WEB-SITE SELECTION IN THE DESERT SPIDER
AGELENOPSIS APERTA
S. E. Riechert, 1976
Oikos 27:311–315

Materials and Methods

The study was conducted at the northern edge of the Malpais Lava Beds, Lincoln Co., New Mexico (Robert M. Shafer Ranch, T6S RIOE. 1636m). Descriptions of the vegetation characteristics of the lava and desert grassland habitats and of the life cycle of *A. aperta* in South Central New Mexico are given elsewhere (Riechert et al., 1973; Riechert. 1974).

Individuals within desert grassland and lava bed study areas were marked in 1971 and 1972 in the following manner. Spiders were captured, etherized, sexed, and paint-marked on the dorsum of the abdomen with one or more colors of a non-toxic, fast drying enamel paint. A leg removal technique was used in immature individuals where paint would be lost upon molting; regenerated legs could be readily distinguished. A total of 222 individuals were marked during the course of the study. The exact locations of occupied webs were mapped and environmental characteristics of web-sites were determined from line-intercept sampling techniques. At each site a line intercept 1 m in length was run in a north-south compass direction bisecting the center of the web sheet with the 50 cm mark positioned at sheet center. Presence and height of plant species, depression depth, and presence of various substrate features were recorded at 10 cm intervals. An individual transect was sampled at a regular distance away from each web (beginning 1 m west of the end of the web transect). This second transect was defined as a nonweb transect regardless

127

of its potential for a past or future web.

The presence or absence of spiders was determined by flushing them from their web funnels each morning between 0800 and 1000, MDST, coinciding with maximum activity (Riechert and Tracy, 1975). Individuals were scored as flushed if they appeared at the funnel entrance when their webs were approached. If a spider did not flush, the web was studied for signs of disuse (i.e., a torn or littered web) and the surrounding area was checked for a new web. When located, new webs were mapped and the occupants flushed to determine their marked status.

HORMONAL CONTROL OF MALE HORN LENGTH DIMORPHISM IN THE DUNG BEETLE *ONTHOPHAGUS TAURUS* (COLEOPTERA: SCARABAEIDAE)

D. J. Emlen and H. F. Nijhout, 1999
Journal of Insect Physiology 45:45–53

Materials and Methods

Rearing beetles

Beetles were collected from horse pastures in Durham County, North Carolina. Pairs of adults were placed in cylindrical buckets (10 cm diameter x 30 cm) three-quarters filled with a moist sand/ soil mixture and provided with unlimited horse manure (methods described more fully in Moczek 1996; Moczek and Emlen in press). Beetles excavated tunnels into the soil, and pulled pieces of dung below ground to fashion dense, cylindrical brood masses, each containing a single egg. Every four days the soil in each of the buckets was sifted, and all brood masses removed and placed in separate soil-filled cups and stored at 28° C and 80% humidity for the duration of larval development.

When larvae molted into the third (final) instar (mean ± st. dev. = 7.8 ± 1.3 days after oviposition) they were transferred to containers that permitted direct observation and removal for weighing. Observation containers were made from plaster of Paris blocks with a 1 cm diameter and 2 cm deep hole drilled into the center. Holes were partially filled with horse manure and covered with microscope coverslips using stopcock grease. Cubes were moistened daily with a spray-mister and kept in sealed, aluminum foil-covered plastic boxes lined with fungicide-treated paper (1% sorbic acid solution).

Ecdysteroid Radioimmune Assay

To characterize the ecdysone titer profile of horned males, hornless males and females of this species, we performed a radioimmune assay for ecdysone using the methods of Borst and O'Connor (1972) and Warren et al (1984). Up to 10µl haemolymph was extracted from each larva by puncturing the abdominal cuticle with a glass needle and collecting the droplet in a capillary tube. Haemolymph samples were immediately blown into 100µl chilled 100% methanol and vortexed. Samples were centrifuged for 3 min at 13,000 RPM, the supernatant transferred to labeled Eppendorf tubes, dried in a vacuum centrifuge, and stored until needed at -20°C. When it was not possible to obtain 10µl from an animal, smaller amounts were used, and final ecdysone counts converted to amount per 10µl.

Methoprene application experiment

To identify the effects of augmented levels of JH on the timing of metamorphosis, as well as its possible effects on male horn determination, we topically applied the JH analog methoprene to developing larvae. Methoprene was dissolved in acetone to a concentration of 10µg/ml. Five microliters of this solution were applied to each larva, achieving an approximate dose of 400µg/g larval weight. Two hundred seventy-one larvae were reared as described above. Larvae of each age were divided randomly among either a methoprene treatment or an acetone-treated control. Methoprene or acetone was administered topically to the dorsum of each larva, immediately behind the head capsule. Doses were applied slowly to prevent loss from run-off or damage to the larvae from solutions entering spiracles. Larvae were observed and weighed daily for the remainder of their development, and the number of days to pupation recorded, as well as the weight, sex and horn morphology of pupae.

129

THE EFFECTS OF ACTINOMYCIN D AND RIBONUCLEASE ON ORAL REGENERATION IN *STENTOR COERULEUS*

G. L. Whitson, 1965
Journal of Experimental Zoology 160 (2):207–214

Materials and Methods

Culturing and cutting techniques. Cultures of *Stentor coeruleus* were obtained from the Carolina Biological Supply Co. Subcultures were maintained at room temperatures (23–25° C) on baked lettuce infusion pH 6.8–7.2 with *Tetrahymena* sp. and *Aerobacter aerogenes* as food organisms.

Cells used for experimental analysis were grown in deep Petri dishes and small depression slides. Refeeding of the stentors was done by daily addition of *Tetrahymena* taken from cultures grown in lettuce infusion which were inoculated daily with fresh bacteria. Both mass cultures and individual isolation lines were maintained during most of this investigation.

Cells for regeneration experiments were pipetted in a large drop of culture fluid onto a small piece of fine-mesh cotton cloth which was attached to a slide with melted paraffin. They were then cut with flame-sharpened tungsten needles kindly provided by S. K. Brahma. A group of tail pieces was obtained by cutting cells during a 15–30 minute interval and placing them in fresh culture medium in spot-plate depressions. In this way it is possible to observe a group of 50–60 tail pieces that will ordinarily regenerate at approximately the same rate.

The use of inhibitors. Actinomycin D (Merck Sharp and Dohme Research Laboratories, West Point, PA) and crystalline ribonuclease (Worthington Biochemical Corp., Freehold, NJ) were used as inhibitors of regeneration. Several concentrations of actinomycin were made up in autoclaved lettuce infusion adjusted to pH 7.0 with 1NNaOH. Ribonuclease in different concentrations was also prepared, and used in either autoclaved lettuce infusion or distilled water adjusted to pH 7.0 with phosphate buffer. Tests with both of these inhibitors were performed on whole cells prior to cutting and on isolated tail pieces at different times during the

130

course of regeneration. Starved and well-fed cells were used. The results were recorded as no inhibition, partial inhibition, or complete inhibition of regeneration.

V. Results

The results of each experiment should be presented clearly, without comment, bias, or interpretation. Figures and tables are often useful but do not substitute for a verbal summary of the findings. The most important features of the figures and tables should be pointed out in this section, with a corresponding reference to the illustration (e.g., see Figure #). We feel illustrative material is so important that we have included a separate chapter on presenting data in tables and figures (Chapter Twelve).

Use a separate paragraph for each major result. Each of these paragraphs should include a **brief** description of the experiment (without repeating the detail of the Materials and Methods), and it should present the results of that experiment. Statements you make about data must be supported with meaningful statistics (see Chapters Seven and Eight). Strive to connect your results together. Guide the reader and point out trends, but be careful not to include any interpretations of your data here. Conclusions about your original hypothesis, as well as comments on the broad implications of your data, should be saved for the Discussion section of the paper.

Here are some sample results sections:

PREY PREFERENCE AND HUNTING HABITAT
SELECTION IN THE BARN OWL
Stephen J. Fast and Harrison W. Ambrose, III, 1976
American Midland Naturalist 96 (2):503–507

Results

In Situation 1, both kinds of rodents were present in both habitats. The owl took significantly more rodents from the field habitat (21) than from the woods (7). The comparison between his prey choices (19 *Microtus*, nine *Peromyscus*) borders on significance.

A field-woods hunting habitat choice with *Peromyscus* was offered in Situation 2. The owl took significantly more mice (11) from the field than from the woods (3).

Situation 3 presented the field-woods choice using only *Microtus*. Again, significantly more rodents were taken from the field (12) than from the woods (3).

In Situation 4, which allowed the owl a choice between the two species in the woods only, the number of *Microtus* eaten (12) was twice the number of *Peromyscus* (6). These results, although suggestive, are not statistically significant, probably due to the small sample size.

The owl was permitted to choose between the two prey species again in Situation 5, this time in the field only. Significantly more *Microtus* (10) were eaten than *Peromyscus* (2).

In Situations 1, 4, and 5, where the owl had the opportunity to choose between the two species of prey, a total of 41 *Microtus* and 17 *Peromyscus* were taken. Situations 1, 2, and 3 gave the owl the field-woods choice; a total of 44 rodents were taken from the field and 13 from the woods. These results, when subjected to the X^2 test, show *Microtus* to be significantly the "preferred" prey item, and the field to be the preferred hunting habitat.

GEOGRAPHIC VARIATION AND NATURAL SELECTION ON A LEAF SHAPE POLYMORPHISM IN THE IVY LEAF MORNING GLORY (*IPOMOEA HEDERACEA*)

K. L. Bright and M. D. Rauser
(unpublished manuscript)

Results

Natural selection in the lobed-only region (Bahama site–1996)

To determine the pattern of natural selection at a lobed only site, the genotypic frequency at a site that naturally had only lobed individuals (Bahama site) was perturbed. We found no difference among genotypes in percent germination or survivorship (see Table 1a). However, we did find fecundity selection for the lobed genotype in this experiment (see Figure 6 and Table 2).

Homozygous lobed individuals produced, on average, about 50% more seeds than homozygous entire individuals, and about 20% more seeds than heterozygotes (Figure 6a and Table 2a). This same general pattern was found with average seed weight (Figure 6b and Table 2b), except both homozygous lobed and heterozygous plants produced heavier seeds than entire plants.

Natural selection in the mixed genotype region (Whiteville and Clinton sites—1997)

At both sites in the mixed genotype portion of the range, we found no difference in germination or survivorship among genotypes (Table 1b and 1c). The pattern of seed production at the Clinton and Whiteville sites was very similar, so both sites were included in a single analysis, with "site" as a factor (no effect of site was found, Table 3). At both sites, heterozygotes had the highest fitness (measured as seed number, Figure 7 and Table 3). None of the pair-wise comparisons (using Scheffe's post-hoc test) were significant. However, according to a contrast between heterozygotes and homozygotes (both lobed and entire), heterozygotes had a significantly higher seed production than both of the homozygotes combined (Table 3).

THE *IN VITRO* MAINTENANCE OF THE APICAL ECTODERMAL RIDGE OF THE CHICK EMBRYO WING BUD: AN ASSAY FOR POLARIZING ACTIVITY
Jeffrey A. MacCabe and Brenda W. Parker, 1975
Developmental Biology 45:349–357

Results
The *in vitro* behavior of the ectoderm and mesoderm of wing bud responding tissue was examined when cultured with tissue from various regions of the wing, both from outside and within the zone of polarizing activity. The results are described below.

Responding Tissue Cultured Alone
In 73 cases responding tissue was placed in culture and incubated for 24, 36, or 48 hr (Table 1, a). In over half of the 24-

hr cultures, the apical ectodermal ridge was not present (Fig. 3a) or was represented only by a remnant at the posterior end of the tissue. In most of the remaining cultures the ridge was present but thin. After 36 hr nearly all of the cultures were without ridge and after 48 hr in culture none had an apical ridge. All of these cultures contained macrophages in the mesenchyme subjacent to the ectoderm (Fig. 3b and c). In most of the cultures the macrophages were scattered along the anteroposterior axis of the responding tissue. In seven of ten cases with a ridge remaining after 24 hr the ridge was confined to the posterior half of the responding tissue and the cell death of the anterior half. After 36 hr in culture the macrophages were scattered along the entire a-p axis of the tissue. The macrophages in the 48-hr cultures were even more scattered, an occasional one even escaping the responding tissue altogether. The data suggest a slight increase in the number of macrophages with increased time in culture.

Responding Tissue Cultured in Contact with Polarizing Tissue

The behavior of the responding tissue *in vitro* was markedly different when cultured in contact with wing tissue from the "zone of polarizing activity" (Table 1, b). In 75 cases polarizing tissue was cultured in contact with either the anterior or posterior end of the responding tissue. In most cases the apical ridge appeared thick and there were no macrophages apparent after 24, 36, or 48 hr of culture (Fig. 4a and b). In the three cases where macrophages were present, they were few. There also ….

Responding Tissue in Contact with Tissue from Outside the Zone of Polarizing Activity

Two sources of wing bud tissue from regions outside the zone of polarizing activity were placed in contact with responding tissue, the anteroproximal corner and the center of the dorsal surface of the wing bud. In addition, tissue from the flank about midway between the leg and wing buds was also used. When the flank or anterior limb tissue was placed in contact with responding tissue the ectodermal ridge of the responding tissue ….

134

COMPETITIVE EXCLUSION
Sue Kirk and Nancy Sagmeister
(students–University of Illinois)

Results

Our results showed that in 27 *E. coli* and *Pseudomonas* crosses
in which the plates had been streaked with equal amounts of
bacteria, 21 plates were dominated by *Pseudomonas* and six were
dominated by *E. coli*.

VI. Discussion

**In this section you evaluate the meaning of your results in
terms of the original question and hypothesis and point out
their biological significance.** The discussion should describe
the significance of your experiment in terms of other work.

Think of the Discussion as a "reverse" funnel. Start narrow
and gradually expand to the broader implications of your study.
The first part of the Discussion should summarize the most
important results and patterns of your specific study. Next, you
should place your result within the context of a summary of other,
relevant studies (with appropriate citations). This section allows
readers to appreciate whether your results are consistent with
expectations from current literature. This is the obvious place to
discuss the potential biological significance of either corroboration
of or deviation from expected patterns. Finally, you may choose
to place your results within an even broader context. Generalize
to other organisms and to the "big picture." You can, and should,
be speculative here, but be careful not to conclude more than
your data permit. Many discussions also include a conclusion at
the end. This is usually a general statement of your conclusions.

Do not attempt to hide unexpected or anomalous results. If
your experiments were designed properly, then all outcomes will
be informative. Unexpected findings often lead to new avenues
of research, which you should make the effort to point out.

Here are some sample Discussion sections:

DISTASTEFULNESS AS A DEFENSE MECHANISM IN *APLYSIA BRASILIANA* (MOLLUSCA: GASTROPODA)
H. W. Ambrose III, R. P. Givens, R. Chen, and K. P. Ambrose, 1979
Marine Behavior and Physiology 6:57–64

Discussion
The results indicate the presence of a presumably distasteful substance which causes gulls to reject significantly more pieces of *Aplysia brasiliana* than fish. None of our observations indicated that *Aplysia brasiliana* was toxic to the gulls, although occasionally a gull which appeared to have swallowed a test piece without tasting it first was seen shaking its head with its mouth open.

As would be expected in a mechanism of defense against predation, the outer, most vulnerable, portions of *Aplysia brasiliana* were the most distasteful. This supports Thompson's suggestion (1960a and b) that the defensive secretions are present in greatest abundance in the areas of skin which would first be encountered by a predator.

It is interesting that any pieces of *Aplysia brasiliana* were eaten at all since few researchers have been able to feed *Aplysia* to other species. Watson (1973) had little luck force-feeding midgut gland to mice; and Ambrose (unpublished data) was unable to feed fresh *Aplysia* to pinfish. However, all of our test pieces were between one and two hours "dead" and Thompson's data (1960a) indicate that dead opisthobranchs lose some of their distasteful properties.

It is perhaps of greater interest that gulls did not discriminate between fresh fish and pieces that had been soaked in the blood, mucus, and ink—a mixture one might expect to be highly distasteful. *Aplysia* ink has traditionally been assumed to play a defensive role (Hyman, 1967; MacMunn, 1895), although it is no longer held that this rather sluggish species makes an escape behind the screen of a cloud of ink. It has been shown that even in the smallest tidepool, the ink does not actually obscure the animal (Kupfermann and Carew, 1974). Other theories about the ink can be found in the literature. Chapman and Fox (1969) suggest that it is merely a waste product of algal metabolism and Tobach et al. (1965) suggest that it communicates information about reproductive states because solitary animals ink less often than those in a group.

It has been our experience that *A. brasiliana* invariably ink when first scooped out of the water in a net and often ink when undergoing a physical examination unless handled very gently. We would like to offer another theory about the ink which does credit it with a defensive role. When Thompson discusses the defensive adaptations of opisthobranchs (1960b), he points out that his data are not consistent with the usual assumption that brightly colored species are normally toxic or distasteful and that cryptically colored species are usually edible. He points out that the fairly cryptic *Aplysia punctata* is as distasteful as the more conspicuously colored opisthobranchs. We would like to suggest that the purple ink of *Aplysia brasiliana* serves as its warning coloration and conclude that this species escapes predation by being a highly distasteful animal and by warning potential predators of this fact with a cloud of purple ink.

ALTERNATIVE TACTICS AND MALE-DIMORPHISM IN THE HORNED BEETLE *ONTHOPHAGUS ACUMINATUS* (COLEOPTERA: SCARABAEIDAE)
Douglas J. Emlen, 1997
Behavioral Ecology and Sociobiology 41:335–341

Discussion
Males in natural populations of the beetle *O. acuminatus* occur in two forms. Populations on Barro Colorado Island, Panama, contain approximately equal numbers of males with a pair of fully developed horns and males with only rudimentary horns or no horns at all (Emlen 1994a, 1997). Here I show that horned and hornless males employ two very different behavioral tactics to encounter and mate with females. Large, horned males guard

entrances to tunnels containing females (Figure 1). Guarding a tunnel enabled a male to mate repeatedly with the female as she provisioned burrows with dung and oviposited eggs. Guarding frequently involved fighting intruding males over tunnel occupancy, and was similar to behavior described for other horned dung beetles living in burrows (*Onthophagus binodis*: Cook 1990; *Onthophagus taurus*: Fabre 1899; Moczek 1996; Emlen D. J., unpublished data; *Phanaeus difformis*: Rasmussen 1994; *Typhoeus typhoeus*: Palmer 1978).

Smaller, hornless males also remained inside tunnels with females when given the opportunity (i.e., when there was no competition from rival males; Figure 2a). However, whenever males competed for access to females (the typical situation in natural populations; Emlen 1994b), hornless males always adopted a non-aggressive alternative tactic that was never employed by the larger, horned males (Figure 2b). Hornless males sneaked into guarded tunnels either by digging side tunnels that intercepted guarded tunnels below ground (Figure 1), or by sliding past guarding males at the tunnel entrance. Sliding past guarding males was similar to sneaking behavior described for hornless males in other horned beetle species (e.g., Rasmussen 1994; Moczek 1996; Moczek and Emlen in preparation), but this is the first characterization of a sneaking tactic involving hornless males digging their own tunnels, and intercepting guarded burrows beneath the soil surface....

Do guarding and sneaking tactics favor horned and hornless male morphologies, respectively? As a first step toward addressing this question, this study measured the effect of natural variation in male horn morphology on male performance at guarding tunnels. In staged contests that controlled for confounding effects of variation in body size, males with relatively longer horns won significantly more fights over tunnel ownership than same-sized males with relatively shorter horns. This suggests that for males large enough to guard tunnels, long horns will be beneficial.

But why should smaller, sneaking males be hornless? One possibility is that males without horns sneak more effectively than males with horns. Horns scrape against tunnel walls as beetles run below ground (Emlen 1994b; Moczek 1996). Sneaking males depend on rapidly entering and exiting tunnels for their reproductive success, and horns may hinder their sneaking performance. Although this remains to be tested for *O. acuminatus*, experiments conducted on the related species *Onthopagus taurus* demonstrated that for same-sized males, males with short horns moved significantly faster inside tunnels than males with longer horns (Moczek 1996; Moczek and Emlen in preparation).

A second possibility is that horns are costly to produce. Relatively small males were not successful at guarding tunnels and presumably derive little benefit from possessing horns. If horns are expensive to produce, then this might favor males able to facultatively omit horn growth whenever developmental conditions preclude the attainment of large body sizes. At least two costs of beetle horns have already been established. First, production of horns significantly extends the development time of *O. taurus* males, and results in increased larval mortality from soil-dwelling nematodes (Hunt and Simmons in press). Second, allocation of developmental resources to horns in both *O. acuminatus* and *O. taurus* results in reduced allocation to other morphological traits, specifically eyes: males with relatively longer horns developed with significantly smaller eyes than males with relatively shorter horns (Emlen and Nijhout in review). Such costs to horn production suggest sneaking males might benefit by not developing horns.

One prerequisite for the maintenance of dimorphism is that organisms experience a fitness tradeoff across environments (Levins 1968; West-Eberhard 1979, 1992; Stearns 1982; Lively 1986b). If animals encounter several discrete environment types, or ecological or behavioral situations, and these different environments favor different morphologies, then distinct morphological alternatives can evolve within a single population —each specialized for one of the different environments. Such fitness tradeoffs have been demonstrated for several dimorphic species. For example, soft and hard seed diets have favored two divergent bill morphologies within populations of African finches

(Smith 1993), and high and low levels of predation have favored alternative shell morphologies in barnacles (Lively 1986a), and spined and spineless morphologies in rotifers (Gilbert and Stemberger 1984) and *Daphnia* (Grant and Bayly 1981; Black and Dodson 1990; Spitze 1992). It is possible that the alternative reproductive tactics characterized in this study produce a similar situation in *O. acuminatus*. If guarding and sneaking behaviors favor horned and hornless male morphologies, respectively, then the reproductive behavior of males may have contributed to the evolution of male horn length dimorphism in this species.

VII. Acknowledgments

Every paper should include a brief acknowledgments section. **In this section, you thank all of the people and organizations that made your study possible**. This typically includes any funding agency that helped pay for your study, as well as, any colleagues who contributed suggestions and/or resources or who helped with the research (e.g., field assistants) or with the writing of the paper (e.g., a friend who edited earlier drafts.)

Here is one example:

I thank B. Allen for use of his hay field, and the North Carolina Department of Agriculture and North Carolina State University for the use of their field stations. The Duke Botany Department graciously allowed me space in their greenhouse and the use of their Herbarium. T. Preuninger and D. Emlen were invaluable field assistants. I received financial support from the Zoology Department at Duke University. I thank D. Emlen, M. Rausher, P. Tiffin, and N. Underwood for their comments on various versions of this manuscript.

VIII. Literature Cited

All published work mentioned in your paper must be listed in this section. You will have saved hours of pain and suffering if you are now working from a neat stack of 3" x 5" index cards, each containing all the bibliographic information needed for a proper citation. List your citations in order of the last name of the first author. In cases of more than two authors, you may use an "et al." in the text of your paper (Jones et al., 1972), but in the literature cited section, all authors must be named. Check every reference against the original to be sure you cite other authors accurately. It is very important to provide credit for the *ideas* of others as well as for experimental findings.

Note: **acceptable sources are those that are peer-reviewed**. Journal articles, edited books, and textbooks that are reviewed by qualified, independent parties before publication are acceptable. However, information obtained from the Internet, while perhaps guiding you in interesting directions, is *not* peer-reviewed, and should not be cited as such. Exceptions to this include technical journals that are accessible online. In these cases, cite the journal article itself, rather than the web address. Here are some sample citations for books, book chapters, and journal articles. Journal titles may be abbreviated according to standard practices (see the *CBE Style Manual* when in doubt).

Literature Cited

Council of Biology Editors, Style Manual Committee. 1994. CBE style manual. 6th ed. Cambridge: Cambridge University Press, 825 p.

Daltry JC, Wüster W, Thorpe RS. 1996. Diet and snake venom evolution. Nature 379:537–540.

Ellner S, Hairston, Jr NG. 1994. Role of overlapping generations in maintaining genetic variation in a fluctuating environment. Am Nat 143:403–417.

Endler JA. 1973. Gene flow and population differentiation. Science 179:243–250.

————. 1977. Geographic Variation, Speciation, and Clines. Princeton (NJ): Princeton Press.

Taylor SE. 1975. Optimal leaf form. In: Gates DM and Schmerl RB, editors. Perspectives of Biophysical Ecology. New York: Springer.

Wyatt R, Antonovics J. 1981. Butterflyweed re-revisited: spatial and temporal patterns of leaf shape variation in *Asclepias tuberosa*. Evolution 35:529–542.

CHAPTER TWELVE

Illustrating Data in Tables and Figures

M ost scientific articles must be augmented with maps indicating the location of a study, diagrams of a study area or apparatus, figures showing graphical or diagrammatic representations of the data, or tables of actual results. Illustrations should be designed with care so that the point of each figure is clear and intuitive. Illustrations provide one of the few outlets for creativity in scientific manuscripts, and well-designed figures enhance the effectiveness of papers susbstantially. In all cases, figures should stand alone—i.e., their point should be obvious to readers independent of the rest of the text (even readers who only skim your paper are likely to study the figures). We first provide examples of several types of illustration and then discuss some tips on making figures especially effective.

I. Sample Illustrations

A. Tables
A great deal of detailed data can be presented in tabular form and merely summarized verbally in the text of an article. Some sample tables follow.

The first table is taken from A. M. Jungreis and G. R. Wyatt, 1972. Sugar release and penetration in insect fat body: relations to regulation of haemolymph trehalose in developing stages of *Hyalophora cecropia*. *Biological Bulletin* 143:367–391.

COMPOSITION OF HAEMOLYMPH OF *Hyalophora cecropia* REARED ON STANDARD OR SPECIAL ARTIFICIAL DIETS

Stage	K	Na	Mg	Ca	Cl	Trehalose (mM)	Osmotic pressure (milli-osmolal)
Larvae							
Mid-fourth instar	28	20	56	9·2	33	35	300
Early fifth instar	35	11	66	12	33	35	307
Early fifth instar, special diet*	44	16	66	9·3	—	33	360
Mid-fifth instar	32	12	69	10	33	44	255
Mid-fifth instar, special diet*	35	9·0	76	6·5	46	41	313
Mature fifth instar (1)†	20	—	40	10	36	44	266
Mature fifth instar (2)	33	5·7	65	8·2	24	45	297
Larval-pupal transformation							
First day of spinning (1)	21	—	40	10	24	35	280
First day of spinning (2)	24	1·4	37	—	22	54	289
Second day of spinning (1)	22	—	45	11	24	31	376
Second day of spinning (2)	30	1·3	54	11	—	44	279
Second day after apolysis (1)‡	33	—	63	12	22	—	311
Second day after apolysis (2)	36	1·3	57	12	29	44	—
Third day after apolysis (1)	31	—	48	12	22	23	287
Third day after apolysis (2)	48	2·3	53	12	22	36	—
Pupae							
Day of ecdysis (1)	39	—	37	11	20	28	373
Day of ecdysis (2)	49	1·6	51	10	23	28	354
+ 1 week 25°C	35	<2	36	11	18	17	343
+ 2 weeks	31	<2	36	10	18	16	480
+ 4 weeks	39	<2	35	11	22	17	584
+ 8 weeks	37	<2	27	11	19	8	>1000§
+ 8 weeks 25°C and 8 weeks 6°C	36	—	46	9·3	—	20	—
Diet, standard	57	19	15	56	35	—	>325
Diet, special	82	19	65	56	110	—	>400

*The standard diet was supplemented with 25 mM MgCl₂ and 25 mM KCl.

†(1) and (2) signify samples from the first and second experimental series, respectively (see Materials and Methods).

‡Apolysis occurred on the third day after the beginning of spinning.

§Values in apparent excess of 1000 milliosmolal represent samples which failed to freeze under the supercooling conditions used, because of their high content of glycerol.

This next table is taken from J. A. McCabe and B. Parker, 1976. Evidence for a gradient of a morphogenetic substance in the developing limb. *Developmental Biology* 54:297–303.

MORPHOGENETIC ACTIVITY IN NORMAL AND EXPERIMENTAL 5-DAY WINGS

Site tested for activity	Number of cultures	Responding tissue after 24 hr of culture			
		Ectodermal ridge			Average number of macrophages per culture
		Thick	Thin	None	
a. Anterior border	35	2	11	22	44.5
b. Middle of limb	37	25	10	2	17.9
c. Posterior border	33	30	3	0	7.3
d. Middle with posterior Mylar barrier	32	4	12	16	49.9
e. Posterior border with posterior Mylar barrier	40	29	11	0	4.0
f. Middle with anterior Mylar barrier	32	11	13	8	16.4
g. Middle with posterior filter barrier	32	11	14	7	19.4
h. Middle, posterior one-third excised	37	4	16	17	45.0
i. Middle, anterior one-third excised	31	3	11	17	33.7
j. Middle, apical one-third excised	33	5	13	15	24.5

This table is taken from S. J. Fast and H. W. Ambrose, III, 1976. Prey preference and hunting habitat selection in the barn owl. *The American Midland Naturalist.* 92 (2):503–507.

The number of rodents eaten of each species in five different combinations of species and habitat types is shown. To the right are the X^2 values for each comparison. * = significance at the .05 level, 1 df. wM and wP = *Microtus* and *Peromyscus* familiarized with the woods habitat. fM and fP = *Microtus* and *Peromyscus* familiarized with the field habitat

	Number of Rodents Eaten				
	Field Habitat		Woods Habitat		
	fM	fP	wM	wP	
SITUATION 1					wM + wP vs. fM + fP
Replicate a	5	2	2	0	X^2 = 7.00*
Replicate b	4	2	0	1	wM + fM vs. wP + fP
Replicate c	5	3	3	1	X^2 = 3.57 N.S.
Total	14	7	5	2	
SITUATION 2					
Replicate a	—	6	-	1	
Replicate b	—	5	-	2	
Total	—	11	-	3	X^2 = 4.57*
SITUATION 3					
Replicate a	5	—	0	-	
Replicate b	7	—	3	-	
Total	12	—	3	-	X^2 = 5.40*
SITUATION 4					
Replicate a	—	—	9	3	
Replicate b	—	—	3	3	
Total	—	—	12	6	X^2 = 2.0 N.S.
SITUATION 5					
	10	2	-	-	
Total	10	2	-	-	X^2 = 5.33*

Our final example of a table is from H. W. Ambrose, III, 1972. The effect of habitat familiarity and toe-clipping on rate of owl predation in *Microtus pennsylvanicus*. *Journal of Mammalogy* 53 (4):909–912.

THE EXPERIMENTAL DESIGN AND RESULTS

	Mice Familiar with environment			
Exp. Test Unit	Number voles available	Predator hours per vole eaten	Toe-clipped voles eaten	Non-clipped voles eaten
1	10	38	4	3
2	10	10	2	3
3	7	24	0	3
4	10	43	4	1
5	8	24	4	1
	Mice unfamiliar with environment			
Exp. Test Unit	Number voles available	Predator hours per vole eaten	Toe-clipped voles eaten	Non-clipped voles eaten
1	10	5	3	2
2	10	4	3	4
3	10	14	2	3
4	10	10	2	3
5	10	6	5	3

B. Figures: Maps and Diagrams

This map is taken from A.C. Echternacht, 1977. A new species of lizard of the genus *Ameiva* (Teiidae) from the Pacific Lowlands and Colombia. *Copeia* 1977 (1):1–7.

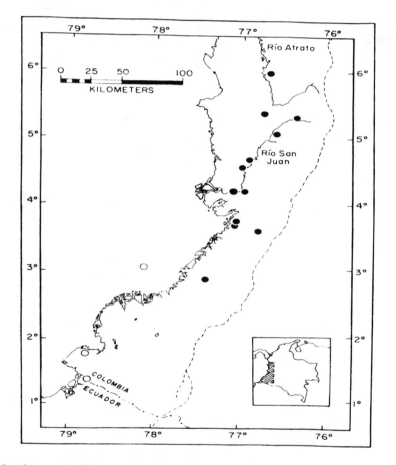

Locality records of Ameiva anomala (solid symbols) and Colombian records of Ameiva bridgesi (hollow symbols). The off-shore locality for bridgesi is Isla Gorgona, Depto. Cauca. The dashed line represents the approximate position of the western cordillera of the Andes.

148

This diagrammatic illustration of potential relationships is taken from S. E. Riechert and C. R. Tracy, 1975. Thermal balance and prey availability: bases for a model relating web-site characteristics to spider reproductive success. *Ecology* 56 (2):265–284.

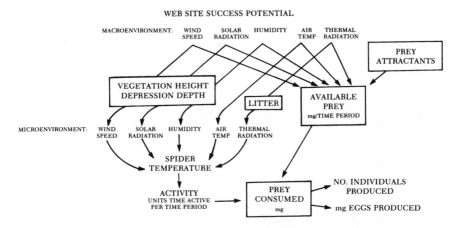

Hypothesized relationship between web-site character and individual reproductive success. Solid lines indicate parameter effects on quantities or functions.

This example of a diagram which illustrates a biological process is taken from G. L. Whitson, 1965. The effects of actinomycin D and ribonuclease on oral regeneration in *Stentor coeruleus*. *Journal of Experimental Zoology* 160 (2):207–214.

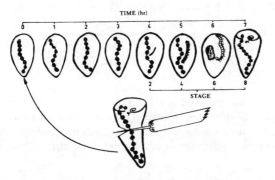

A diagram of oral regeneration in bisected stentors. Regeneration occurs in 6-8 hours. The stages shown are those given by Tartar ('61). The beaded macronucleus coalesces and renodulates late during oral regeneration.

C. Figures: Graphs

Graphs can give a quick, visual illustration of significant trends in experimental data. This first graph is taken from R.M. Bagby, 1974. Time course of isotonic contraction in single cells and muscle strips from *Bufo marinus* stomach. *American Journal of Physiology* 227 (4):789–793.

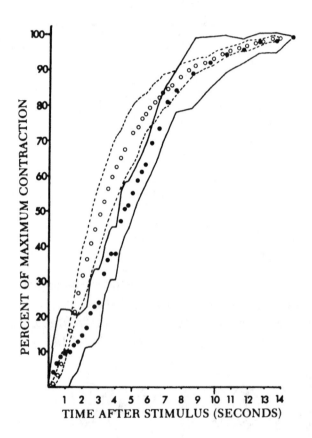

Comparison of mean time course of samples of cells (open circles) with mean time course of muscle strips$_c$ (solid circles). Original data are rescaled so that individual responses use their own final lengths as 100% maximum contraction. Dashed lines and solid lines outline 95% confidence limits for cells and strips$_c$, respectively.

This next graph is taken from A. M. Jungreis, P. Jatlow and G. R. Wyatt, 1974. Regulation of Trehalose synthesis in the silkmoth *Hylophora cecropia*: the role of magnesium in the fat body. *Journal of Experimental Zoology* 187 (1):41–45.

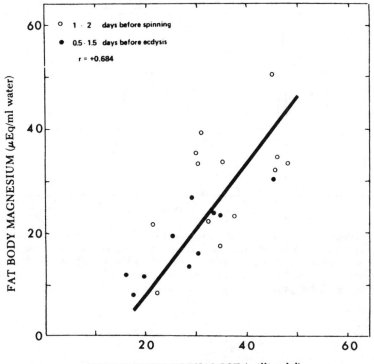

Linear correlation analysis between the concentrations of trehalose in hemolymph and magnesium in fat body in larval and pharate pupal stages of development.

This graph was taken from G. L. Whitson, J. G. Green, A. A. Francis and D. D. Willis, 1967. Cyclic changes in the alcohol-soluble carbohydrates in synchronized *Tetrahymena*. *Journal of Cellular Physiology* 70 (2):169–177.

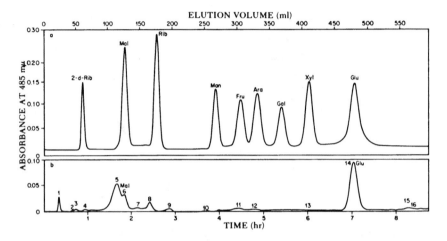

Chromatograms of samples of (a) 9-components of an 18-component standard containing 0.5 micromoles of each sugar and (b) 1 ml extract of synchronized cells obtained 75 minutes after EHS. Peaks are numbered arbitrarily for identification. Note that peak 6 (maltose) and peak 14 (glucose) are two of the major peaks identifiable by parallel chromatography with the standard (a).

The next two graphs are taken from S.E. Riechert and C.R. Tracy, 1975. Thermal balance and prey availability: bases for a model relating web-site characteristics to spider reproductive success. *Ecology* 56 (2):265–284.

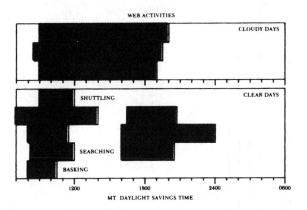

Periodicity of various web activities for clear and cloudy days. Bars represent time periods in which 90% of each of the designated activities were observed. Vertical lines represent the medians of these activities.

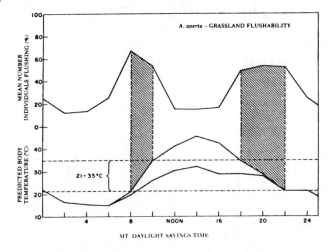

Graph of percent of spiders active on the sheet with time of day on the mixed-grassland study area in midsummer (July and August) imposed on predicted spider temperature under these conditions and assuming a web-over-litter substrate. Barred area under flushability curve represents time periods during which over 50% of the individuals were active. Stippled area represents range of spider temperatures, exact temperature dependent upon amount of exposure to solar radiation. Upper boundary of predicted temperature curve signifies spider temperature if in full sunlight. Lower boundary signifies spider temperature if in full shade. Area enclosed by dashed lines represents body temperature range within which over 50% of the spiders are active.

The last two graphs are from H.W. Ambrose III, 1973. An experimental study of some factors affecting the spatial and temporal activity of *Microtus pennsylvanicus*. *Journal of Mammalogy* 54 (1):79–110.

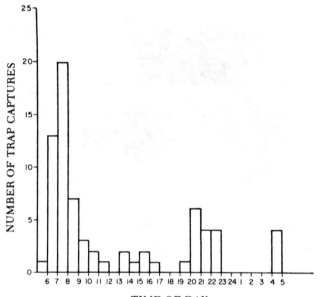

TIME OF DAY

Relationship of trap captures to time of day for all study voles.

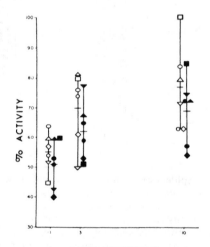

MICE PER PEN

Mean percentage activity of individual male (closed symbols) and female (open symbols) *Microtus* at three population densities. Individuals are indicated by differently shaped symbols.

154

II. Tips On Making Figures More Effective

Some general guidelines may be helpful in designing figures for your paper. Ask yourself "What is this figure (or table) supposed to accomplish?" Every illustration should make one, and generally only one, point. Once you are clear on what, precisely, each illustration will say, then decide on the most effective means of conveying this information.

Should you use a table or a figure? Tables are helpful if you have repetitive data to present, or if the exact numerical values of your data are important (e.g., for comparison with related studies). Tables are often more difficult to read than figures, and they are difficult for editors to print. Use them only when necessary. Every table should have a clear title that explains its point, and "like" elements within the table should be arranged in columns, not rows (so they are read down the table, rather than across).

Figures are preferable to tables when the important result involves trends or patterns, rather than exact numerical values. If your experiment included treatments, and what matters is the difference between treatments and controls, then figures are likely to be much more effective at getting this point across than tables.

Once you have decided on a type of figure (scatter plot, bar chart, etc.), there are many ways to make the figure more accessible to readers. Computer graphics programs give you control over the appearance of any given graph. This includes the capacity to change the fonts and font sizes of the axes labels, the range of the axes, the size and shape (and even color) of the symbols used for the points, and whether curves/regression lines are solid or dashed. Play with these options, and find arrangements that make your point especially clear. It is often helpful to import graphs into illustration programs (like Macromedia Freehand or Adobe Illustrator) for still greater flexibility (e.g., the addition of pictures or line drawings).

Here is an example of a figure "before" and "after." The plot shows data from an experiment where a hormone was applied to developing beetle larvae. Since the important point concerns the response of treated animals relative to controls (i.e., a trend, rather than exact numerical values), a figure was used instead of a table.

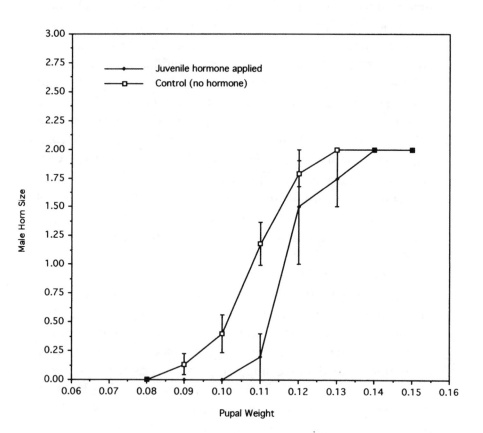

As plotted by the original graphics program (in this case, Cricket Graph for Macintosh), this figure contains extraneous and unnecessary information, as well as wasted space. Part of making illustrations easier to understand involves culling unnecessary information. Here, most of the numerical values on the axes are not needed, and by removing half of them we make the remaining numbers easier to see. In addition, by changing the range of values included in the axes, we can expand our data to fill a larger percentage of the figure "box" (i.e., we remove all of the space at the top of the previous figure).

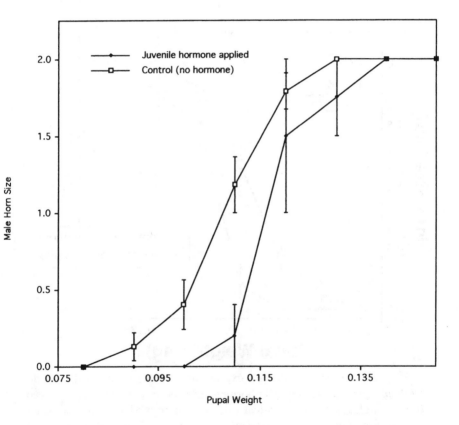

We can improve this illustration further by making all of the important information larger and more conspicuous. For this final version, font sizes were enlarged considerably. Symbol points and lines were enlarged, and an added level of redundancy reinforced the visual comparison of "treatment" versus "control." Now, not only are the symbols "open" or "closed," but the fitted curves are also open (dashed), or closed (solid). This type of subtle redundancy can make figures extremely effective. Finally, a line drawing was added to illustrate the study organism and the focal trait (horns in the beetle are shaded in black).

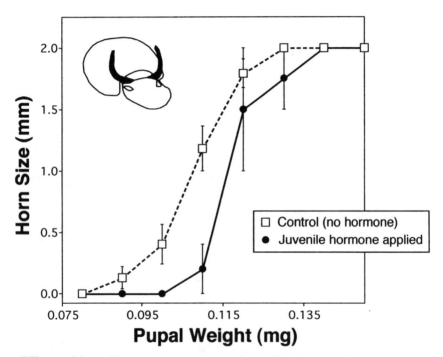

Effects of juvenile hormone on expression of male horns in the beetle *Onthophagus taurus*. Juvenile hormone applied to male *O. taurus* larvae (closed circles) shifted the critical weight for horn production as compared with acetone-treated control larvae (open squares). Bars indicate standard errors.

Our second example shows how leaf shape affects plant reproduction. As plotted by the graphics program, the bar chart has font sizes that are much too small to be read easily. This graph also does not include enough information on the Y-axis (we need more numbers to appreciate the scale), and it includes unnecessary ticks along the axes.

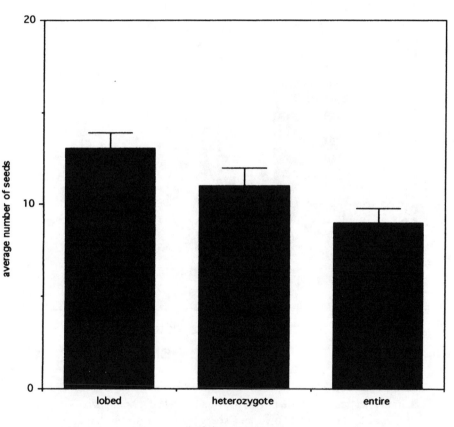

We can improve the effectiveness of this figure by fixing the above problems, and also by adding redundancy to the point of the graph. As in the example above, subtle redundancies can make the point of the figure much easier for a reader to grasp.

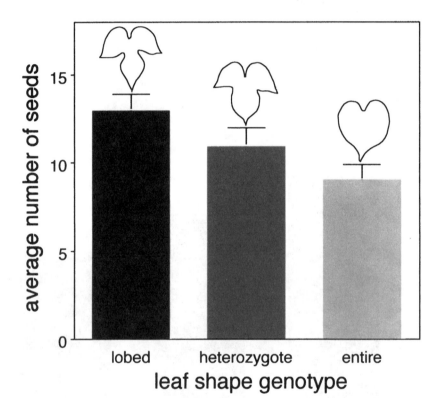

Natural selection on leaf shape genotype in the ivyleaf morning glory (*Ipomoea hederacea*). Plants of known leaf shape genotype were planted in equal frequencies into a wild field. Leaf shape genotype significantly affected seed production, with lobed-leaved plants producing more seeds than either heterozygous plants or entire-leaved plants (one-factor ANOVA: F=3.90; p<0.01).

In this case, the three treatment categories represent a gradation from highly indented (lobed) leaves to not-indented (entire) leaves. This gradient is now illustrated by drawings of each of the leaf types. This presents, in a different way, the same information as the labels on the X-axis. The same gradient is also illustrated by the pattern of shading of the bars, with the darkest bar being the most-lobed leaf type, and correspondingly lighter bars for the less-lobed leaf forms. Thus, three separate techniques (axis labels, line-drawings and bar-shading) *all* show that the data categories represent a gradient of leaf shapes, reinforcing the point of the figure.

CHAPTER THIRTEEN

Problems and Pitfalls In Writing a Scientific Paper

In this chapter we will touch upon a few common technical difficulties associated with writing the scientific paper. For more detailed help we recommend the *CBE Style Manual* and another Council of Biology Editor's publication: *Scientific Writing for Graduate Students*, edited by F. Peter Woodford; and *Writing to Learn Science* by Randy Moore.

I. Citing other work

Factual information taken from published sources should be documented whether or not you use an actual quotation. Footnotes are seldom used in scientific writing. Generally, reference in the text is made only to the author's or authors' names and date of publication—the full bibliographic information appears only in the Literature Cited section. Both the name and date can go inside parentheses if the name is not actually part of your sentence. For example:

> *A large part of the natural diet of barn owls consists of* Microtus *(Wallace 1948; Craighead and Craighead 1956)....*

If the author's name is intended to be part of the text, only the date goes in the parentheses:
Metzgar (1967) has shown that ....

When the date itself is important to your text, you omit the parentheses:

> *As early as 1967, Metzgar showed that ....*

If there are more than two authors, the citation in the text can be condensed to an "et al." form; however, the full citation must appear in the Literature Cited.

> *...theory of population regulation (Hairston et al. 1960).*
> *Hairston et al. (1960) have suggested that ....*

II. Polishing your paper

A direct, concise and precise style can be achieved only through practice. Always allow time for at least two drafts of a paper. When you have put all of your ideas down, read through them carefully, consider each word and eliminate extraneous verbiage. Rewrite faulty or misleading grammatical constructions. It is also helpful to have other people read your drafts and mark where they have difficulty understanding your point. These are the sections you will need to work on.

In this chapter we will give some typical examples of sloppy, unpolished writing taken again from student papers. As you read the examples, try to think of how to improve them before you read the suggested corrections. This exercise will help you spot similar difficulties in your own papers.

III. Wordy, awkward constructions

> *It is thought that possibly the reason why Hawaii does not fit the area-species curve of the islands around it is because of its recentness.*

It is possible Hawaii does not fit the area-species curve of the islands around it because of its geological recentness.

> *However, when dealing with small microorganisms, identifying them apart from each other is quite difficult.*

However, microorganisms are difficult to distinguish.

> *The results may be restated to say that the directed movement response of female* Acheta domesticus *to a tone versus silence, a cricket song versus silence, and a cricket song versus a tone, is not significant in terms of which one the insects will move toward.*

Movement of female *Acheta domesticus* was not significantly directed in any of the three choice experiments—tone versus silence, cricket song versus silence, or cricket song versus tone.

By our research we are attempting to test whether the minority effect is valid.
In this study we test the validity of the minority effect.

How species may compete with each other has been based on models of how crowded animals of the same kind may so compete among themselves that their population growths are curbed and their crowds are controlled.
Mathematical models suggest that competition limits growth rate and density in a variety of species. [Note: this sentence must cite the authors of the models.]

The question we are asking is: Using three different habitats, what is the diversity and abundance of animals within each habitat, and how do they differ?
How do the diversity and abundance of animals differ among the three different habitats?

It appears by looking at Graph A that ....
Graph A shows that ....

Often wordy sentences can be improved by removing the passive voice.

In an investigation by Hutchinson of the animals of an intertidal rocky shore, it was noticed that the adults of two species of barnacles occupied two ....
Hutchinson (1965) noticed that two species of barnacles on an intertidal rock shore....

164

Dangling participles and other constructions in which a sentence somehow gets the wrong subject can be embarrassing.

Using sterile technique, the bacteria ....
After injecting the hormone, the frog ....
By studying competition, ecology can better understand....
Numerous experiments within the last ten years have reported the mutagenic, carcinogenic, or chromosome damaging effects of ....
[Here the authors really say, but cannot mean, that bacteria use sterile technique, frogs inject the hormone, ecology studies competition, and experiments do their own reporting!]

Try to use verbs instead of abstract nouns. Expressions such as "was accomplished," "was performed," and "were attained" should be warning signals. These phrases are pompous.

The calculation was performed
We calculated ...

The measurement was attained
We measured ...

Often a paper lacks precision because a word or phrase has not been used properly.

The results of the experiment demonstrate that fish cannot be taught certain behavior, conditioned, and thus fish learn behavior pattern by instinct.
[Here the student makes a rather grandiose conclusion based on his failure to condition a fish and jumps unwittingly into the foggy area of learning versus instinct by carelessly implying that things can be "learned by instinct."]

When two species are competing for the same space, usually one will overpopulate the other.
[Presumably the student meant simply that one species might outnumber the other, but permitting "overpopulate" to have a direct object suggests that possibly he means "eliminate."]

Out of the 45 plates we streaked, Pseudomonas *dominated the bacterial cultures in 33 of the plates.*
[The sinister political implications of "dominating a culture" are unintentional. Presumably "predominated in the culture" is implied.]

... the males began with aggressive stridulations towards the females.
[Can a cricket stridulate <u>towards</u> another cricket? Probably not. Toward is the correct word.]

Supposedly smaller islands hold smaller populations and are therefore subject to faster extinction.
[Does this mean that the islands are supposedly smaller but we are not sure? Or that smaller islands supposedly hold smaller ....]

Fish have commonly been known to associate certain smells, colors, or sounds to mean different things.
[They may well associate these things <u>with</u> other things but that is all.]

Avoid incomplete sentences.

On our first day we obtained two stocks. One of white apricot mutant fruit flies and the other of red-eyed, wild-type fruit flies.

On our first day we obtained two stocks—one of white ....

or

On our first day we obtained two stocks. One was white ....

Beware of ambiguous antecedents.

Onion bulbs were partially submerged in water with the root end just below the surface. Three toothpicks were affixed to the onions to hold them in place. These were grown for three days.
[Could the toothpicks have grown any in that short time?]

If the number of protozoans on the islands have been reduced during dispersal to remote places, then they have successfully attained the distance effect.
[Have the protozoans or the remote places attained the distance effect?]

It has been found that the group diversities of these organisms differ with the habitats, which can be attributed to the influence of different physical characteristics in the habitats.
[It would be clearer to say that the fact that the group diversities differ can be attributed ..., and avoid the confusion over "habitats which ..."]

Do not change the form of the verb mid-sentence.

We observed which flies were male and which were female by first knocking them out with ether and then, using a magnifying glass, we looked at the back of the flies.
[first knocking, then looking]

Do not give a plural subject a singular verb form.

Traits of immunity enables the species to continue because ....
Traits enable ....

As a final kindness to your readers, break up noun clusters and stacked modifiers

One cricket-call cassette tape recording was obtained on a ....

We obtained one cassette tape recording of a cricket call ....

As a final suggestion, consider this poem.

> If you've got a thought that's happy,
>   Boil it down.
> Make it short, and crisp, and snappy—
>   Boil it down.
> When your brain its coin has minted,
> Down the page your pen has sprinted,
> If you want your effort printed,
>   Boil it down.
>   —Anonymous

Bibliography and Suggested Sources

PHILOSOPHY OF SCIENCE

Ingle JD. 1958. *Principles of Research in Biology and Medicine.* Philadelphia: JB Lippincott Co.

Moore JA. 1993. *Science as a Way of Knowing: The Foundations of Modern Biology.* Cambridge (MA): Harvard University Press.

National Academy of Sciences/National Academy of Engineering/Institute of Medicine. 1995. *On Being a Scientist: Responsible Conduct in Research.* Washington (DC): National Academy Press.

Platt JR. 1964. Strong Inference. *Science* 146:347–353.

STATISTICS

Brown FL, JR Amos, OG Mink. 1995. *Statistical Concepts, A Basic Program.* 2nd ed. New York: Harper and Row.

Guilford JP. 1978. *Fundamental Statistics in Psychology and Education.* McGraw-Hill Series in Psychology. 6th ed. New York: McGraw-Hill Book Co.

Rohlf FJ. 1994. *Statistical Tables.* 3rd ed. San Francisco: WH Freeman Co.

Siegel S. 1988. *Nonparametric Statistics for the Behavioral Sciences.* 2nd ed. New York: McGraw-Hill Book Co.

Sokal RR. 1995. *Biometry: The Principles and Practices of Statistics in Biological Research.* 3rd ed. New York: Freeman.

Steel RGD and JH Torrie. 1980. *Principles and Procedures of Statistics: A Biometrical Approach.* 2nd ed. New York: McGraw-Hill Book Co.

Terrace H and S Parker. 1971. *Psychological Statistics.* 7 vols. San Rafael (CA): Individual Learning Systems, Inc.

Zar JH. 1984. *Biostatistical Analysis.* 2nd ed. Englewood Cliffs (NJ): Prentice-Hall, Inc.

WRITING A SCIENTIFIC PAPER

Council of Biology Editors, Committee on Form and Style. 1994. *Scientific Style and Format: The CBE Manual for Authors, Editors, and Publishers.* 6th ed. Cambridge and New York: Cambridge University Press.

Gopen GD and JA Swan. 1990. The science of scientific writing. *American Scientist* 78: 550–558.

Gubanich A. 1977. Writing the Scientific Paper in the Investigative Lab. *The American Biology Teacher.* January 1977.

Moore R. 1997. *Writing to Learn Science.* Fort Worth: Saunders College Publishing.

Woodford PF, editor. 1986. *Scientific Writing for Graduate Students: A Manual on the Teaching of Scientific Writing.* Bethesda (MD): Council of Biology Editors.

THE INVESTIGATORY LABORATORY

Biological Sciences Curriculum Study. 1976. *Research Problems in Biology, Investigations for Students.* Series 1, 2 and 3. 2nd ed. New York: Oxford University Press.

The Carolina Biological Supply Company publishes a newsletter, *Carolina Tips*, which is a good source of information on laboratory techniques.

Morholt E. 1986. *A Sourcebook for the Biological Sciences.* 3rd ed. New York: Harcourt Brace Janovich. (The appendix to this book has an excellent bibliography of sources for teaching biology and equipping laboratories.)

Thornton JW, editor. 1972. *The Laboratory: A Place to Investigate.* Commission on Undergraduate Education in the Biological Sciences. Publ. No. 33. (Supported by a grant from the National Science Foundation to the American Institute of Biological Sciences.)

Turtox/Cambosco, MacMillan Science Co., a supply company for scientific equipment, publishes a newsletter, *Turtox News*, with useful laboratory hints and techniques.

Appendix

S tudents at any level can be prepared to perform original research. The essential prerequisite is a careful, gradual introduction to the scientific method. We offer you our general plan for the introductory weeks of any investigatory laboratory course, not because it is the only one that works but because we have found it successful.

We have frequently heard derogatory comments such as, "Only a few honors students can succeed in this sort of program," or "Non-majors are not capable of independent research," or "It's a fine idea but it won't work with freshmen," but we have never found them to be true. It is true that some inadequate educational systems manage to stifle creativity and independence, but in the proper atmosphere they can be quickly revitalized and nourished. Students rarely exceed your expectations, so you must believe in them, encourage them, and guide them toward a scientific way of thinking.

In the introductory weeks outlined below, students learn to identify problems, ask appropriate questions about them, restate these questions in terms of null and alternative hypotheses, and design methods of answering these questions, figuring out what information would be needed and how to collect it.

I. First week in the laboratory

At the first session, in introductory remarks, the instructor points out that the primary methods of gathering scientific information are observation and experimentation. Before students learn to experimentally manipulate phenomena, they should consider the process of observation itself. All knowledge is subjective, all information about the external world must be acquired by and filtered through the human nervous system. The best we, as scientists, can hope for is to learn to use our powers of observation with as little human bias and subjective interpretation as possible. This is a matter of practice.

The students are then given specimens of plant and animal material. Typically these are leaves and crickets. (We usually have three habitat boxes, one containing six males, one with six females, and one with three of each sex, which the class can take turns examining.) The students are asked to relax (nothing will be graded), take their time, make observations, and write down any data they consider significant. Later the data are collected, compared, and discussed as a group.

Most students collect good visual data but fewer record observations of sound, smell or taste. Many of their observations are colored by anthropomorphic assumptions. (For example, the female cricket is often assumed to be the male because the ovipositor is seen as a penis.) This should lead naturally into a discussion of bias and objectivity.

In **the second laboratory period of the first week**, students have more practice collecting data. The instructor drops two balls (tennis and Ping-Pong) simultaneously in front of a vertical yardstick or meterstick, and the students, armed with tally counters and stopwatches, are asked to collect any data they think are appropriate. The drop is repeated as often as necessary until the students are satisfied they have completed their observations. Once again the data are collected, compared, and discussed. Students are not graded on their work.

In the discussion, it should quickly become apparent that different people collect different data because they have different questions in mind. Students may ask which ball bounces higher, or how many bounces does the ball make before it stops bouncing, or what is the average height of the first bounce, etc. The point to be underscored is that, to be meaningful, data must be aimed at answering a particular question.

After the discussion of the bouncing ball situation has subsided, the class should be urged to start asking questions about anything they can think of and then to restate these questions into answerable form. If no questions are forthcoming, students should mill about, look at various displays in the laboratory, look out the window, look at each other, etc. Any kind of question they might answer with simple observations or experimentation would be suitable for the purposes of discussion. Sample questions might be: What make of car passes the window of the lab most frequently? What is the average age (in months, perhaps) of the students in this class? How many of the students here are first children in their families? This exercise should lead to a discussion of the different kinds of data and different scales of measurement described in Chapter Three of this handbook.

Materials needed in the first week:
leaves
crickets (in habitat boxes if possible)
stopwatches
tally counters
two balls, visually distinct
a meter or yardstick

II. Second week in the laboratory

By the end of this week, students, working in pairs, should have performed four "mini-experiments" on simple systems such as dice, cards, spinners, coins, balls and metersticks, or other phenomena of their choice. They should be able to come up with questions, refine the questions into answerable form, and restate them in terms of null and alternative hypotheses. In addition, they should be able to determine what kind of data would answer their questions, how much data would be needed, and which statistical test would be appropriate to determine the statistical significance of their results. Naturally, at the beginning, the students will need a lot of help. Class discussions about the various questions and data collected are particularly helpful.

At the opening laboratory session of week two, there should be a short review of asking questions. This time the questions should be restated in terms of null and alternative hypotheses.

As pairs of students find questions they think they would like to use for their mini-experiments, they should be encouraged to begin. Research pairs should start by writing a brief outline of their research proposal. This should contain the question being asked, the null and alternative hypotheses, an outline of the methods and materials, and mention of the statistical treatment to be applied to the results. These written proposals should be checked by the instructor before the team begins its data collection.

If students are having trouble finding questions that differ from examples already used by the instructor, they should be encouraged to think of comparative questions. For example, if all they can think of when looking at a coin is to ask whether the coin is fair (an example they would have gotten from the handbook), they should be moved toward questions such as the following: Can Charlie flip more heads than Sue? Do the results of tosses caught on the hand differ from those of tosses which land on the floor?

The students should write brief reports of the results of their experiments, including the actual data collected, its analysis, and its statistical significance. Mini-experiments should take approximately one hour from proposal to report. These experiments should be graded in terms of clarity of thought and experimental design. It is not important that the results be statistically significant, although conclusions drawn should be consistent with the original question asked and the actual data collected.

Materials needed in the second week:
stopwatches
tally counters
coins, dice, spinners, playing cards or other gambling
    paraphernalia
balls
metersticks

III. Third week in the laboratory

This week, working in pairs, the students design and perform an experiment of their choosing concerning the human blood circulatory system. Usually the experimental questions concern the rate of heartbeat, the blood pressure, or the body temperature (or some combination of these) because these aspects of the system are easily measured.

Fellow students and any others they can persuade to cooperate are available as experimental subjects. Frequently faculty, secretaries, janitors and other staff are railroaded into volunteering as experimental subjects.

The research pairs have one week to design their experiment, have their proposal accepted by the instructor (instructors should be careful to consider the safety of the project), gather their data, analyze it, and write it up. Preparation, data collection, analysis, and writing usually take place in the laboratory.

Materials needed for the third week:
stethoscopes
sphygmomanometers
thermometers
stopwatches
people willing to be experimental subjects
handouts of background information

(We have prepared a description of the human heart and related phenomena such as pulse and blood pressure, mentioning the roles of hormones and the sympathetic and parasympathetic nervous systems. We also have several "how to" sheets: How to measure the heartbeat rate. How to measure body temperature. How to measure blood pressure.)

IV. Fourth week in the laboratory

This week, working in pairs, the students design and perform an experiment relating to germination in the seeds of the alfalfa plant, *Medicago sativa*. Once again the students must come up with their own questions, formulate their hypotheses, and have research proposals approved by their instructors. Most phases of this study are carried out in the lab. A written research report is required and is usually prepared outside class.

Materials needed in the fourth week:

seeds of alfalfa, *Medicago sativa*

petri dishes

paper toweling (for providing a suitable, damp environment for germination)

tissue paper layers or aluminum foil (to cover petri dishes with materials allowing differing degrees of light penetration)

thermometers (for locating and recording different temperature environments, for example)

potting soil or sand (for varying the depth of germination and other phenomena associated with depth such as oxygen availability)

salt, fertilizer, dextrose, sucrose, glucose, and possibly other substances (to vary the growth environment in terms of its acidity, nutrients, etc.)

rulers (to measure root production as an indication of germination)

temperature controlling devices such as refrigerators, heaters, etc.

handouts discussing dormancy and germination and suggesting possible environmental factors which might make seeds come out of dormancy

V. Fifth week in the laboratory

The students have one more week-long experiment to perform and write up. This time there is much more freedom of choice in the topic. They are asked to inspect the laboratory supplies such as petri dishes, glassware, plastic boxes, lamps, hot plates, ice, aluminum foil, colored paper and cellophane, stopwatches, thermometers, respirometers, sand, soil, and various chemicals, making a mental inventory of what is available. Then they are given an organism list of specimens which the staff has on hand for student use. A typical list might be: pill bugs, roaches, planaria, snails, flour beetles, crickets, meal worms, seeds and algae. Research teams select their organism and think up any experiment they can using the lab supplies or things they manufacture themselves. Again, research proposals must be approved and final reports submitted.

A suggestion sheet is usually provided to help them find suitable research questions. Typical topics from the suggestion sheet are:

Habitat selection. Do some animals actively seek out certain temperatures, pH, light conditions, colors, etc?

Distribution. How do the animals distribute themselves in their environment? Are they in clumps, randomly distributed, or uniformly distributed? What factors seem to influence the distribution process?

Orientation. What directional cues are the animals using as they move about?

Social Behavior. (Particularly suitable for crickets) Are the crickets territorial? Do they have a social hierarchy and, if so, what factors seem to influence social rank?

Respiration. How do different environmental factors (such as temperature, light, etc.) affect the behavior (or germination) of the organism? A handout is available on how to determine the rate of respiration for air-breathing animals and seeds.

VI. From week six to the end of the term

During the final weeks of the term, the students design, execute, and write up one major scientific experiment. This time the topic is geared to mesh with the material being studied in the course. For example, if the students are learning genetics, they might do anything from a classical *Drosophila* experiment to a survey at shopping centers to see if human populations are in a Hardy-Weinberg equilibrium. In other quarters or semesters of the course, the research projects are also coordinated with the broad areas being studied.

For the major projects, lists of possible research topics and organisms are provided along with a few handouts of techniques and background information. This time, in preparing the full (no longer in an outline) research proposal, students must search the literature in the library to inform themselves about their topics and locate published techniques where needed.

Research proposals should be carefully examined to be sure the students understand their questions and how they will answer them. Both grandiose and overly simple experiments should be discouraged so that no student is working either above or below his capability.

Research results should be reported in the form of a full, formal, scientific paper with the necessary figures and tables carefully prepared. Once again, the grading should consider the experimental logic, clarity of presentation, accurate analysis of data, and intelligent conclusions based upon results. Whether or not the experiments actually succeed in providing a statistically significant answer to the question posed should be inconsequential. In all cases, the scientific approach used, not the actual "success" of the experiment, is paramount. Needless to say, the research performed does not have to be original as long as it results from a question which is genuinely original to the student. The work will be valuable whether it finds something previously unknown or merely supports other similar findings in the field.

VII. Why we teach this way

By this method of teaching, we hope to give each student the ability to think critically and analytically, to ask pertinent questions, and to design tests that would answer them. This way of thinking and method of problem solving should foster curiosity and ingenuity and should give each student a taste of the excitement of science. The Investigatory Laboratory method of teaching is a truly self-paced, self-directed, and particularly self-rewarding experience. Armed with the scientific method, the student can discover and uncover facts for himself—he or she holds the key to unlock the secrets of the universe.

Glossary/Index

abscissa – The abscissa is the horizontal axis of a graph. In a graphic representation of data such as a frequency histogram, the vertical side, or ordinate, designates the frequency of the values recorded and the abscissa is used for the different scores or values of the parameter being measured. The lowest value occurs at the left end of the abscissa; see page 16.

abstract – An abstract is a one or two paragraph condensation of the main elements in a scientific article. For sample abstracts from published scientific papers, see pages 120–122. Because an abstract can give the reader a quick impression of the material published, abstracts are often published separately by services which provide a guide to current literature; see pages 114–117.

alpha – The "alpha level" (or α) is an arbitrary level of risk. When using a test to determine whether you reject your null hypothesis, you knowingly set the level of risk, such as a 5% chance, that your test will lead you to accidentally reject a true null hypothesis (a Type I error); see page 46.

ANOVA – ANOVA, or the analysis of variance, is a parametric statistical test of differences between means of more than two samples; see pages 51–57.

asymptotic – This term is used to describe the relationship between the tails of a normal curve and the baseline (abscissa) of a graph. The tails of the curve of a normal distribution are asymptotic to the baseline, or abscissa, receding indefinitely and never actually touching the baseline because the perfect normal curve represents an infinite number of cases; see page 26.

bimodal distribution – When a group of ordered data, plotted on a frequency histogram, reveals two peaks of frequently occurring values, it is described as having a bimodal distribution; see page 20.

Chi-Square (χ^2) – The Chi-Square One-Sample Test for Goodness of Fit is a test for differences between distributions; see pages 92–95. The Chi-Square (χ^2) Test of Independence between Two or More Samples is a test for independence between two or more frequency distributions of nominal data; see pages 96–98.

conclusion – For a discussion of drawing conclusions based on the results of statistical tests, see page 48. In writing a scientific paper, the conclusions you draw from your experimental results are presented in the section entitled "discussion." For sample discussion sections, see pages 135–140.

continuous data – Continuous data are variables from a scale such as length or time which is a continuum. Although there will be actual gaps between the items measured and recorded, there is always the theoretical possibility of additional data points between them. Unlike the measurement of numbers of children, discrete data with no possible child-and-a-half fractions falling between two consecutive readings, in continuous data fractional intermediate values are always possible; see page 13.

correlation – Variable factors which are related to each other, such as shoe size and height are said to have a correlation. Correlation does not necessarily imply a cause-and-effect relationship between the two variables. In the example mentioned, because taller people usually have larger shoe sizes, the factors are said to have a positive correlation. Altitude and air pressure have a negative correlation because, as altitude increases, air pressure decreases; see page 40.

discrete data – Discrete data consist of items that are separate, whole units. The numbers of children in sample families would be plotted as discrete, whole number units. There cannot be a fractional increment on the scale between 1 child and 2, or between 2 and 3 children, etc.; see page 12.

discussion – In a formal, scientific publication, the discussion is the last section of the text before the "literature cited." The discussion normally presents the conclusions drawn from experimental results and analyzes these findings in terms of the question posed and the current state of knowledge about the topic under consideration. For samples, see pages 135–140.

dispersion – The dispersion of data is its spread on either side of a central tendency. Dispersion can be measured in terms of its range (the distance between the highest and lowest reading) or its variance, a measure of how clumped the data are; see pages 24–31.

distribution – see frequency distribution.

F Test – The F test is a test for difference between variances; see pages 85–91.

frequency distribution – When data such as scores, or other measurements of some variable population, are arranged in order of magnitude, this distribution reveals the frequency of occurrence of different values. When the ordered scores are represented in a table or graph which indicates the number of times a particular score or value occurs in the group, the data are represented in a frequency distribution. Frequency distributions of data are often displayed in a frequency histogram; see page 16.

frequency histogram – The frequency histogram is a graphic representation of data indicating how frequently different values occur. In the graph, the frequency of occurrence is plotted on the vertical axis (ordinate) with frequency increasing with the height of the axis. The data being plotted, whether discrete or continuous, are distributed along the horizontal axis (abscissa). In the case of data which can be ordered, the direction of increase is from left to right. In a histogram, the scores are represented by rectangular boxes of differing heights (frequencies) over the appropriate points on the abscissa; see page 16.

Friedman Test – The Friedman Test is a non-parametric test of differences between means; it is used to test for significant differences between the responses of several matched samples (paired samples) exposed to three or more treatments; see pages 58–62.

hypothesis – A hypothesis is a possible explanation for a phenomenon. Scientific investigation involves the testing of predictions based on hypotheses; see page 1. See also null hypothesis.

interval scale – There are two scales of measurement for continuous data—the interval and the ratio scale. Both scales of measurement indicate the distance between items of continuous data, but the units of measurement on the interval scale are not fixed by an absolute zero point. Zero degrees Fahrenheit is an arbitrary endpoint which does not signify the absolute absence of heat. January 1 is an arbitrary starting point but can hardly be considered the beginning of time; see page 13.

introduction – In a scientific paper, the introduction presents the question being asked, or hypothesis being tested, and justifies the experiment by indicating what is currently known about the topic under investigation and why this particular question is of interest. Findings by other authors are usually mentioned in this part of the paper. For examples of introductions, see pages 122–126.

Kruskal-Wallis – The Kruskal-Wallis Test is a non-parametric test of differences between means; see pages 63–67.

linear regression – See regression analysis.

literature cited – The literature cited is the last formal section of a scientific paper. It is a bibliographic list of published material actually mentioned in the article and is presented in alphabetical order based on the surname of the first author of each entry. For a brief discussion, see page 140. For help in deciding what bibliographic data to collect when doing a literature search, see page 104.

mean – The mean (often written as $\overline{X}$), a calculation of central tendency in a frequency distribution, is the numerical average. The mean is computed by dividing the sum of the scores by the number of scores; see page 22.

median – The median, a measure of central tendency, is the middlemost value in a distribution of data. To find the median, the data scores or values recorded must be organized in a progressive sequence. In the case of an even number of readings, an exact median can be calculated by averaging the two central data points (adding them together and dividing the sum by two). See page 22.

methods and materials – In a formal scientific paper, the section entitled methods or methods and materials, describes the materials and procedures used in the investigation being reported in sufficient detail that another scientist could repeat the experiment. Previously published techniques can be cited and not repeated in detail. Even though this section is supplied so that an experiment could be repeated for purposes of verification, it should be written in the past tense, telling what was done not what the reader should do. See pages 126–131 for examples.

mode – The mode, a measurement of central tendency, is the most frequently occurring value in a distribution of data. It is possible for a distribution to have more than one mode. The mode is easily revealed if the data points are ordered sequentially or plotted on a frequency histogram; see page 23.

nominal scale – Discrete data, "measured" on a nominal scale, are data assorted into named groups. Nominal measurement usually involves counting items in the separate categories; see page 12. See ranking scale for comparison.

normal distribution – Large groups of quantitative data usually approximate the theoretically normal frequency distribution characterized by a symmetrical, "bell-shaped" curve with a mean, median, and mode of the same value. This curve, a frequency histogram smoothed of its steps because all theoretical points are illustrated, has an equal number of scores on either side of the central axis (or mean). The tails of the curve are asymptotic to the baseline (abscissa) infinitely receding and approaching but never touching it. There are two points on the normal curve where the direction of curve changes from convex to concave. These are the points of inflection. See pages 21–23 for more information.

null hypothesis – The null hypothesis (H_0) is the hypothesis of no difference. All statistical tests are designed to determine whether you have sufficient reason to reject your null hypothesis. Since there is no such thing as positive proof, scientific advances are made by the rejection of null hypotheses. If your data indicate a statistically significant reason to reject your null hypothesis, for example that the mean of sample A is equal to the mean of sample B (let us say at the .05 alpha level of confidence that you have not made a mistake), then you can accept one of your alternative hypotheses (that the mean of A is greater than or less than the mean of B). This type of logical basis for a statistical test is typically written in shorthand:
H_0: $\overline{X}A = \overline{X}B$ H_1: $\overline{X}A > \overline{X}B$ or H_2 : $\overline{X}A < \overline{X}B$. See pages 34–36.

ordinate – The vertical axis of a graphic presentation of data. In a frequency histogram, the ordinate indicates the frequency of occurrence of the values measured, with frequency increasing with the height of the ordinate; see page 16.

ordinal scale – An ordinal scale, or ranking scale, of measurement is used to organize items of discrete data in terms of some meaningful relationship to each other. The nominal categories can be ranked according to some aspect. For example, categories of makes of cars can be organized in terms of their increasing weight along an ordinal or ranking scale (and along an abscissa); see page 12.

parametric statistics – Parametric statistical tests are limited to statistically normal distributions of data whose data points or individual observations are independent of each other and distributed on the same, continuous scale of measurement. In general, if these criteria can be met, parametric statistics should be selected over non-parametric tests because they are more "powerful" tests, giving you a greater ability to reject your null hypothesis; see page 45.

parameters – Parameters are aspects of a population of data such as the mean or the range; see page 11.

points of inflection – In the bell-shaped curve characteristic of a normal frequency distribution, the points of inflection are those points on either side of the mean where the direction of the curve changes from convex to concave. Theoretical lines drawn from the points of inflection to the abscissa mark off the standard deviation on either side of the mean; see page 26.

population – The term population is used in statistical tests to describe any group of similar kinds of things being measured; see page 11.

range – The range of a distribution is a measurement of its dispersion around the mean. It is the distance between the lowest and the highest reading; see page 24.

ranking data – To rank a series of values, you assign the rank of 1 to the lowest, 2 to the next highest, and so forth. See the illustration in the Spearman's Rank Correlation Test on page 81.

ranking scale – The ranking scale is the same as the ordinal scale of measurement. It is a meaningful order or ranking imposed on categories of discrete, nominal data; see page 12.

Rank Sum Test – The Rank Sum Test is a non-parametric test of differences between means; see pages 68–71.

ratio scale – The ratio scale is a scale of measurement for continuous data. Unlike the interval scale, the ratio scale has a true zero point although the units of measurement are arbitrary. The majority of the data you collect will be measured on a ratio scale where the possibility of no growth, no time elapsed, no distance traveled, etc. will exist; see page 13.

regression analysis – regression analysis deals with the study of dependent relationships between correlated variables. There is a brief discussion of this kind of correlation on page 41. The statistical test for simple linear regression begins on page 72.

Sign Test – The Sign Test is a non-parametric test of differences between means; see pages 78–80.

Skewed Distribution – When data plotted on a frequency histogram are not evenly distributed on either side of a central high point, they are skewed. In positively skewed data, the mean is located to the right of the majority of readings. In a negatively skewed distribution, the mean is located to the left of most of the values plotted; see page 18.

Spearman's Rank Correlation – The Spearman's Rank Correlation is a non-parametric test of relationships which may be used to determine whether two variables are correlated; see pages 81–84.

standard deviation – The standard deviation (written as "s" or σ), is a standard unit of measurement of deviation or distance from the mean along the abscissa of a frequency distribution. This unit is a measurement away from the mean point on the baseline (to the left and/or to the right) to the point where a theoretical, perpendicular line from the point of inflection crosses the abscissa. Because the tails of the normal curve are asymptotic to the baseline, an infinite number of standard deviations from the mean exist, but three standard deviations from the mean, on either side (a total of 6), account for 99.74 percent of all the values, by definition. One standard deviation on either side of the mean (a total of 2) includes 68.2 percent of the possible data values, and two standard deviations (a total of 4) include 95.44 percent; see pages 26–31.

t Test – The t Test is a parametric test for differences between means of independent samples; see pages 85–91.

Type I and Type II errors – When using a statistical test to evaluate a null hypothesis, two types of errors can occur. A Type I error is the rejection of a true null hypothesis. If you set alpha at the .05 level of confidence, the chances are 1 out of 20 that your statistical test will accidentally allow you to reject a null hypothesis that actually was true. The Type II error is the failure to reject a false null hypothesis. A lower alpha level will decrease your chance that you will fail to reject a null hypothesis that really is false. (The .05 alpha level is the normal compromise position between these two risks.) See pages 46–48.

uniform distribution – In a uniform distribution of data, the frequency of occurrence of each value is the same and the bell-shaped normal curve is flattened to a straight line; see page 17.

variance – The variance is a measurement for describing the dispersion of data around the mean. By definition, it is the square of the standard deviation. It is written as s^2. Although it is a useful parameter for certain statistical tests, it is not plotted on most displays of data where the standard deviation should be used because of its size; see pages 25 and 31.

χ^2 – See Chi-Square.